Lab Manual
to Accompany
Science of Earth Systems

Lab Manual
to Accompany
Science of Earth Systems

2nd Edition

Stephen D. Butz

THOMSON

™

DELMAR LEARNING

Australia   Canada   Mexico   Singapore   Spain   United Kingdom   United States

© 2007 Thomson Delmar Learning, a part of the Thomson Corporation.
Thomson, the Star logo, and Delmar Learning are trademarks used herein under license.

For more information contact Thomson Delmar Learning,
5 Maxwell Drive, Clifton Park, NY 12065.

Printed in United States of America
1 2 3 4 5 XXX 10 09 08 07

Or find us on the World Wide Web at http://www.delmarlearning.com

Library of Congress Catalog Card Number: 2006038076

ISBN-10: 1-4180-4124-6   ISBN-13: 978-1-4180-4124-3

Contents

Preface

The Science of Earth Systems Lab Manual, 2nd Edition is a practical, hands-on lab manual designed to complement the topics presented in *The Science of Earth Systems,* 2nd Edition text-book. The 55 activities in this lab manual are practical, easy to administer, and time-tested in the classroom. They provide new, innovative ways to teach the subjects that together make up the science of Earth systems. This lab manual will also allow students to explore, through hands-on investigation, the interactions that occur in the living and nonliving world. Many of the labs presented in *The Science of Earth Systems Lab Manual* are designed to be extremely flexible, and require easily obtained materials that are available to all educators. *The Science of Earth Systems Lab Manual* can also be adapted for use with existing educational technologies. Many of the labs give you the option of incorporating the use of computers and educational technology if it is available to your students. If not, the labs can still be used in the traditional classroom setting. *The Science of Earth Systems Lab Manual* introduces students to the processes that occur on our Earth by dividing the planet into five unique spheres. These include the exosphere, lithosphere, atmosphere, hydrosphere, and biosphere. Together, these five spheres cover all the principal interactions between the Earth's physical and biological properties that make our planet unique. Any educator who needs to teach secondary students about the basic functions of the Earth through hands-on investigation will find this lab manual extremely useful.

Lab Safety Guidelines

1. Follow all instructions clearly.

2. Identify the location, and be familiar with, the procedures for using all safety equipment in the laboratory or classroom.

3. Never eat or drink while performing laboratory experiments.

4. Report all injuries or accidents to your instructor immediately.

5. No running or horseplay in the lab.

6. Do not perform unauthorized experiments and use all equipment only as directed.

7. Return equipment and supplies where you got them, and dispose of materials as instructed.

8. Keep your work area clean and uncluttered.

9. Wear appropriate clothing while performing a lab. Beware of loose sleeves, long hair, and open-toed shoes.

10. Wear safety glasses at all times when instructed.

11. Be careful with hot materials or chemicals of any kind.

12. Be careful when using fire or a heat source of any kind.

13. Allow materials and containers to cool thoroughly before handling them.

14. Never smell chemicals or breathe in fumes of any kind.

15. Report spills or breaks to your instructor immediately.

16. Handle all specimens with care.

17. Never throw anything.

18. Never taste any chemical or substance of any kind.

19. Use care when using sharp or pointed objects.

20. Never pick up any broken glass with your hands.

21. Never operate electrical equipment near water or with wet hands.

22. Wash your hands thoroughly after performing any lab activity.

Lab 1
Measurement and Percent Deviation

Purpose

The purpose of this lab is to have you become proficient in using scientific instruments for precise measurement. The ability to make precise measurements and observations is a fundamental aspect of scientific investigation.

Materials

meter sticks	250-ml beakers
balances or scales	thermometers
rock samples	empty cardboard box

Procedure A

Use the instruments provided to determine the following measurements. Make sure to record the resulting measurement (including correct units) and the instrument you used to make the following measurements.

1. Length of the classroom

2. Height of your desktop

3. Area of your desktop

4. Area of one floor tile

5. Volume of the cardboard box

6. Mass of the rock sample

7. Mass of 250 ml of water

8. Temperature of water from faucet after running for 10 seconds

9. The number of seconds in this period

Procedure B

Using your recorded measurements and the accepted values provided by your instructor, determine the percent deviation in Table 1–1 using the following formula:

$$\text{Percent Deviation} = \frac{\text{difference between measured value and accepted value}}{\text{accepted value}} \times 100$$

		TABLE 1–1 Percent Deviation		
	Recorded Measurement	Accepted Value	Calculations	Percent Deviation
1				
2				
3				
4				
5				
6				
7				
8				
9				

Conclusions

1. What are the two fundamental parts of all measurements?

2. Explain why scientific instruments are used to make measurements.

3. Explain why percent deviation is used when making scientific measurements.

LAB 2
Scientific Notation

Purpose

The purpose of this lab is to use scientific notation, also called exponential notation, to record various measurements commonly used in Earth systems science. Scientific notation is used in science to easily represent extremely large or very small numbers. It makes it easier to write and manipulate these numbers.

Materials

worksheets for Procedures A and B

Procedure A

In this part of the lab activity, you will convert measurements from their scientific notation form into their common form. Record your answers below each number on the worksheet.

1. Half-life of carbon 14 = 5.7×10^3 years

2. Half-life of uranium = 4.5×10^9 years

3. Average radius of the Earth = 6.37×10^3 kilometers

4. Average radius of the Sun = 6.96×10^5 kilometers

5. Wavelength of gamma radiation = 1×10^{-13} meters

6. Wavelength of X-ray radiation = 1×10^{-10} meters

7. Wavelength of ultraviolet radiation = 1×10^{-5} meters

8. Wavelength of blue light = 4.3×10^{-5} meters

9. Wavelength of green light = 5.3×10^{-5} meters

10. Wavelength of red light $= 7.0 \times 10^{-5}$ meters

11. One milligram $= 1 \times 10^{-3}$ meters

Procedure B

In part B, you will convert measurements from their common form into scientific notation. Record your answers below each number on the worksheet.

1. Total surface area of the Earth = 196,950,711 square miles

2. Distance from the Earth to the Sun = 92,960,117 miles

3. Distance from the Earth to the Moon = 238,866 miles

4. Height of the highest point on Earth above sea level (Mt. Everest) = 29,022 feet

5. Deepest point in the ocean (Marianas Trench) = 36,198 feet

6. Diameter of a particle of sand = 0.006 centimeters

7. Diameter of a particle of silt = 0.0004 centimeters

8. Diameter of a particle of clay = 0.00006 centimeters

Conclusions

1. What does a negative exponent mean when used in scientific notation?

2. What does a positive exponent mean when used in scientific notation?

3. Explain why scientific notation is used for measurements in science.

LAB 3
The Electromagnetic Spectrum

Purpose

The purpose of this lab is to have you become familiar with the different forms of electromagnetic radiation within the electromagnetic spectrum, and to identify the relationship between wavelength and energy within the electromagnetic spectrum.

Materials

graph paper   colored pencils

Procedure

Using the data provided in Table 3–1, create a multiple line chart that displays the relative wavelength of seven portions of the electromagnetic spectrum. The *x*-axis of your chart will be labeled "Relative Wavelength" and should be numbered from 1–35. The *y*-axis of your chart will be labeled "Relative Wave Height" and should be numbered from 1–30. Plot each relative wavelength and wave height data point on your graph for each portion of the electromagnetic spectrum in a different colored pencil. Connect the data points and label each portion of the electromagnetic spectrum using a colored pencil.

TABLE 3–1 Wavelengths and the Electromagnetic Spectrum Graph

Relative Wave-length	Gamma Rays	Relative Wave Height			Infra-Red	Micro-waves	Radio Waves
		X-Rays	Ultra-violet	Visible Light			
1	26	22	18	14	10	6	2
2	29						
3	26	25					
4	29		21				
5	26	22		17			
6	29				13		
7	26	25	18			9	
8	29						5
9	26	22		14			
10	29		21				
11	26	25			10		
12	29						
13	26	22	18	17		6	
14	29						
15	26	25					2
16	29		21		13		
17	26	22		14			
18	29						
19	26	25	18			9	
20	29						
21	26	22		17	10		
22	29		21				5
23	26	25					
24	29						
25	26	22	18	14		6	
26	29				13		
27	26	25					
28	29		21				
29	26	22		17			2
30	29						
31	26	25	18		10	9	
32	29						
33	26	22		14			
34	29		21				
35	26	25					

x–axis = "Relative wavelength" (1–35)

y–axis = "Relative wave height" (1–30)

Conclusions

1. Describe the relationship between wavelength and energy within the electromagnetic spectrum.

2. What portion of the electromagnetic spectrum has the shortest wavelength?

3. Which portion of the electromagnetic spectrum has the longest wavelength?

4. List the seven colors of the visible light portion of the electromagnetic spectrum from shortest to longest wavelength.

LAB 4
Earth Coordinates—Latitude and Longitude

Purpose

The purpose of this lab is to have you become familiar with the use of the latitude and longitude coordinate system to locate exact positions on the Earth. This lab will also introduce you to the relationship between longitude location and time.

Materials

world maps or globes

Procedure A—Flight of Scientific Discovery

Imagine you are the leader of a scientific expedition that has chartered its own airplane to travel the world in search of scientific discovery. Your starting point can be anywhere in the United States, and you must travel the world in search of the scientific places of interest listed in Table 4–1. From your starting point, you must fly to a destination were you can observe the place of interest listed. You must record the distance you flew to get there, the latitude and longitude location of the place, and the name of the location. You must also plot your course in colored pencil on the map of the world (Figure 4–1) provided in the lab.

Procedure B—Longitude and Time

Complete the following steps.

1. Calculate the number of degrees of longitude the Earth moves in one hour by using the following information. Show your work.
   The Earth makes one complete rotation on its axis (360 degrees) in 24 hours.

TABLE 4–1   Latitude and Longitude Flight of Scientific Discovery

Place of Interest	Latitude/Longitude Location	Location Name	Distance Traveled
A Glacier			
A Tropical Volcano			
The Highest Mountain in the World			
Greenwich, England			
A Coral Reef			
A Mid-Latitude Volcano			
A Large Lake			
A Canal			
An Island			
A River Delta			

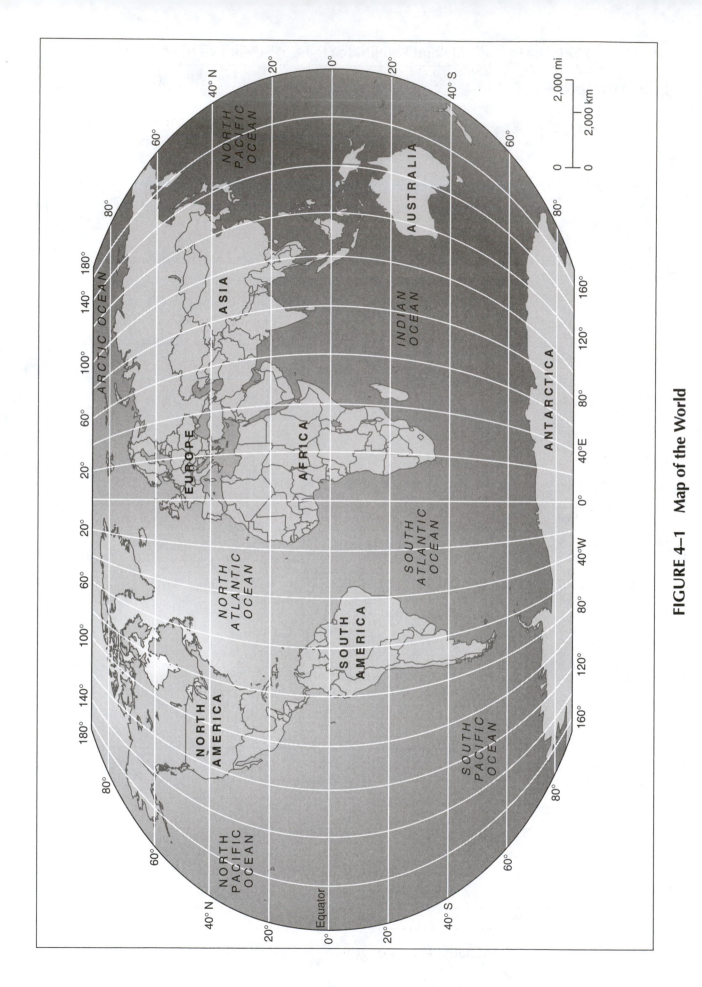

FIGURE 4–1   Map of the World

2. Using your answer from the above question, calculate the following longitude locations. (Hint: if your local time is behind Greenwich Mean Time (GMT), you are west longitude; if your local time is ahead of GMT, you are east longitude.)

a. It is 12:00 noon GMT and 11:00 A.M. local time. What is your longitude location?

b. It is 12:00 noon GMT and 1:00 P.M. local time. What is your longitude location?

c. It is 7:00 A.M. local time and 12:00 noon GMT. What is your longitude location?

d. It is 4:00 P.M. local time and 1:00 P.M. GMT. What is your longitude location?

e. It is 12:00 midnight GMT and 6:00 P.M. local time. What is your longitude location?

Procedure C

Use the diagram in Figure 4–2 showing longitude lines spaced 15 degrees apart to determine the time of day for each location.

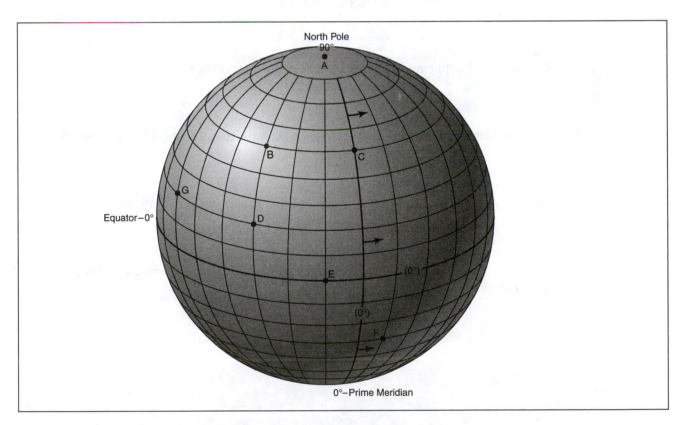

FIGURE 4–2 Longitude Line Diagram

1. If GMT is 12:00 noon, what time is it at location B?

2. If it is 6:00 P.M. at location D, what time is it at location F?

3. How many hours time difference exist between locations G and E?

4. If it is 11:00 A.M. at location G, what time is it at location F?

5. If it is 3:00 P.M. at location D, what time is it at location B?

Conclusions

1. What do lines of latitude measure on the Earth?

2. What do lines of longitude measure on the Earth?

3. How far does the Earth travel in degrees of longitude each hour?

4. What information do you need to determine your longitude location on the Earth?

5. Explain how you can determine if your location is east or west longitude.

6. Explain why it took a long time for an accurate method of determining longitude location on the Earth to be developed.

LAB 5
Motions of the Moon

Purpose

The purpose of this lab is to have you plot the azimuth and altitude locations of the Moon as it moves across the nighttime sky. In this lab, you will learn about the shape, location, and rate of movement of the Moon's apparent path.

Materials

graph paper or computer spreadsheet application

Procedure

Use the data in Table 5–1 that shows the azimuth location and altitude of the Moon on the night of October 31, 2001, to create a line chart of the path of the Moon. The *x*-axis will be labeled "Azimuth in Degrees" and the *y*-axis will be labeled "Altitude in Degrees."

Conclusions

1. Use your completed line chart to answer the following questions.
    a. Describe the shape of the Moon's path through the sky.

    b. In which direction does the Moon rise?

    c. In which direction does the Moon set?

    d. What was the highest altitude the Moon reached during the night?

    e. Approximately how many hours did it take for the Moon to move across the sky from moonrise to moonset?

    f. Calculate the rate of movement for the Moon in degrees of azimuth per hour. Show your work.

13

g. What direction was the moon at its highest altitude?

h. Describe the similarities in the apparent paths that the Sun and Moon make across the sky.

TABLE 5–1 Path of the Earth's Moon on the Night of October 31, 2001, from 4:53 P.M. to 6:34 A.M.	
Azimuth (degrees)	Altitude (degrees)
78	0
83	5
88	11
93	16
98	22
103	27
109	32
115	37
123	42
131	46
140	50
151	53
163	55
176	56
189	56
202	55
213	52
224	49
233	45
241	41
248	36
254	31
259	26
265	21
270	15
275	10
280	5
285	0

LAB 6
Life Cycle of Stars

Purpose

The purpose of this lab is to introduce you to the different aspects of a star's life cycle. All stars go through a similar sequence of events throughout their lives. Understanding this sequence allows you to recognize the unique characteristics of different stars.

Materials

computers with Internet access

Procedure A

Type in the following Web site address into your computer's Web browser: <http://cse.ssl.berkeley.edu/bmendez/ay10/2000/cycle/cycle.html>

Use the "Stellar Life Cycle" Web site to complete the data in Table 6–1 on the life cycle of stars.

TABLE 6–1 Life Cycle of Stars	
Life-Cycle Stage	Main Characteristics

15

Procedure B

Using the information from the Web site and the data table from Procedure A, fill in the flow charts in Figure 6–1 with the correct sequences of each star's life cycle.

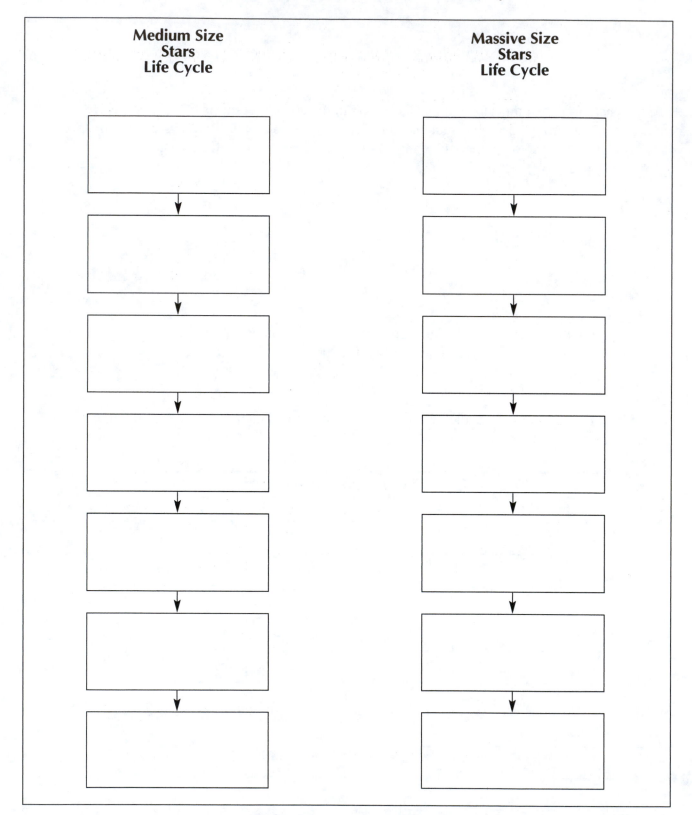

FIGURE 6–1   Star Life Cycles

Conclusions

1. Describe some of the physical characteristics of a star that can be used to determine which part of its life cycle it may be in.

2. List the general stages that together make up the life cycle of a medium-sized star.

LAB 7
Star Classification

Purpose

The purpose of this lab is to have you identify the main characteristics used to classify stars, and the five main types of stars. You will also become familiar with the use of the Hertzsprung-Russell diagram used for the classification of stars.

Materials

graph paper   colored pencils

Procedure

Using the data from Table 7–1 and the blank Hertzsprung-Russell diagram in Figure 7–1, plot the position of each star using its approximate temperature and luminosity. Label each star's name next to its data point on the chart.

TABLE 7–1 Star Temperature and Luminosity		
Star	Temperature (°C)	Luminosity (compared to the Sun)
Rigel	14,000	50,000
Betelgeuse	3,500	12,000
Polaris	6,500	1,000
Aldebaran	4,000	100
Barnard's Star	3,000	0.002
Alpha Centauri A	6,000	3
Sun	6,000	1
Procyon B	6,600	0.01
Sirius B	8,300	0.01
Sirius A	10,000	20
Vega	9,700	60
Tau Ceti	5,000	0.5
Alpha Centauri B	4,300	0.3
Regulus	12,300	300
Achernar	16,700	1,000
Spica	19,700	800
Beta Centauri	21,000	1,200

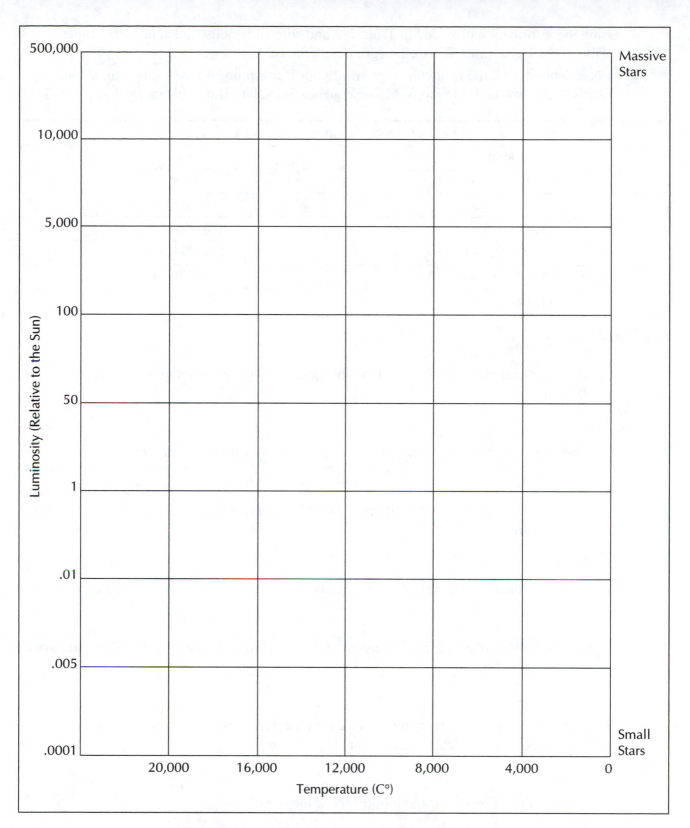

FIGURE 7–1   The Hertzsprung-Russell Diagram

Using the information provided in Table 7–2 and colored pencils, shade in each temperature region of the Hertzsprung-Russell diagram with the correct star color.

In bold letters, label the following regions of your Hertzsprung-Russell diagram: White Dwarfs, Red Dwarfs, Red Giants, Main Sequence Stars, and Blue Supergiants.

TABLE 7–2 Star Temperature and Color	
Star Temperature (°C)	Star Color
2,000–4,000	Red
4,000–5,000	Orange
5,000–6,000	Yellow
6,000–7,500	Pale Yellow
7,500–11,000	White
11,000–22,000	Pale Blue

Conclusions

1. As a star changes color from red to blue, describe what happens to its surface temperature.

2. The Hertzsprung-Russell diagram classifies stars by which four properties?

3. A main sequence star that is 10,000 times more luminous than the Sun most likely has a temperature of:

4. A main sequence star that has a luminosity of 100 is most likely to be what color?

5. A white dwarf star with a temperature of approximately 10,000 degrees C would have a luminosity of:

6. A massive star with a temperature of 20,000 degrees C and a luminosity of nearly 1,000,000 would be classified as what type of star?

7. What is the temperature and luminosity of the sun?

8. The sun is brighter than which two star types?

LAB 8
Life Cycle of the Sun

Purpose

The purpose of this activity is to have you observe the changes in the temperature, absolute magnitude, and other observable characteristics of two different types of stars as they go through their life cycles. The absolute magnitude is a measure of how bright a star would appear if it was approximately 32 light years away from the Earth. One of the stars you will observe will be a medium-sized star similar to our own Sun, and the other star will be a massive star over 100 times the size of the Sun.

Materials

colored pencils
ruler

Procedure A

1. Use the data on the absolute magnitude and temperature for a Sun-sized star in Table 8–1, to plot the location of each life-cycle stage on the blank HR diagram (Figure 8–1A). Plot each of the life cycles in the appropriate color that the star would be based on its temperature. Next to each point, label the life-cycle stage.

TABLE 8–1		
Sun-Sized Star		
Life Cycle Stage	Temperature (°C)	Absolute Magnitude
Proto Star	3,000	0.4
Main Sequence	6,000	4.9
Red Giant	2,500	–5.0
White Dwarf	8,000	13.0
Massive Star		
Proto Star	7,000	2.5
Main Sequence	12,000	–1.0
Super Red Giant	9,000	–7.0
Neutron Star?	17,000	7.0

2. Using a green colored pencil, draw an arrow from the nebula location to the proto star stage. Continue to draw a green arrow connecting each preceding life-cycle stage to the next. When you are finished, make a key on your HR diagram that shows the green line representing the life cycle stages of a Sun-sized star.

3. Next, use the data on the absolute magnitude and temperature for a massive star in Table 8–1 to plot the location of each life-cycle stage on the blank HR diagram (Figure 8–1B). Plot each of the life cycles in the correct color that the star would be based on its temperature. Next to each point, label the life-cycle stage.

4. Using a purple colored pencil, draw an arrow from the nebula location to the proto star stage. Continue to draw a purple arrow connecting each preceding life-cycle stage to the next stage. When you are finished, make a key on your HR diagram that shows the purple line representing the life cycle of a massive star that is over 100 times larger than the Sun.

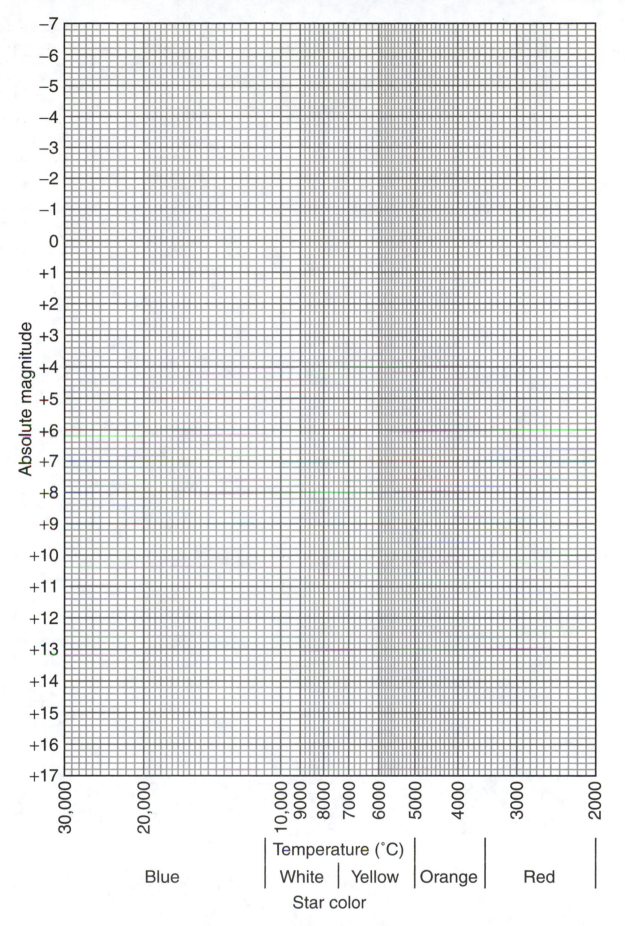

FIGURE 8–1A

23

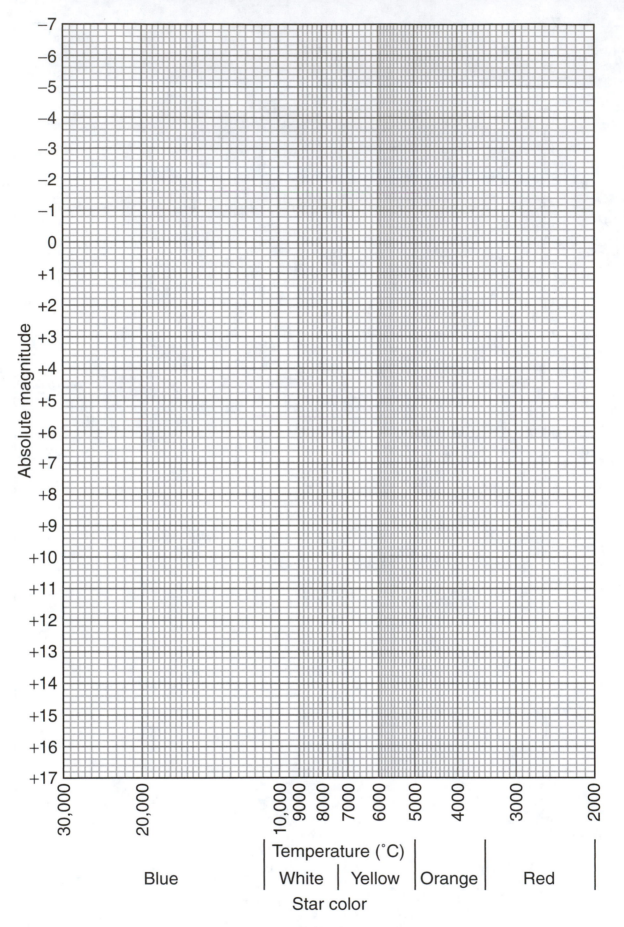

FIGURE 8–1B

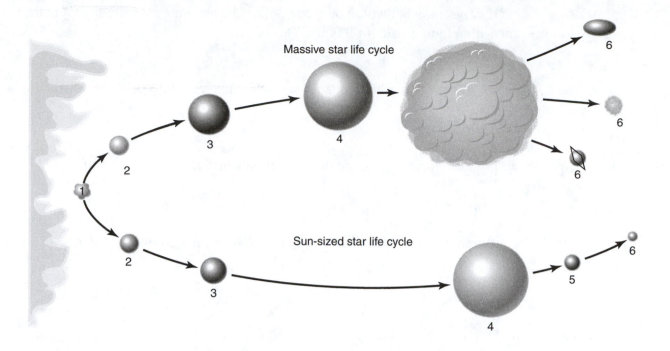

FIGURE 8–2

Procedure B

1. Label the six different stages for a Sun-sized star as shown in Figure 8–2.

2. Label the eight different stages for a massive star as shown in Figure 8–2.

3. Use the correct colored pencil to color in each specific stage as shown in Figure 8–2.

Conclusions

1. What changes occur in the temperature and size of a Sun-sized star as it goes through its life cycle?

2. What are the main differences that can occur in the life cycles of a Sun-sized star and a massive star?

3. What color are the hottest stars?

4. What color are the coolest stars?

5. Using your HR diagram, what would be the approximate absolute magnitude of a main sequence star with a temperature of 9,000°C?

6. If a red giant is cooler than a main sequence star, what makes it have a higher absolute magnitude?

7. After a super nova explosion, what are the three things that can result in the life cycle of a massive star?

8. What is the difference between the luminosity and absolute magnitude of a star?

LAB 9
Relative Sizes of the Planets

Purpose

The purpose of this lab is to have you use a computer spreadsheet application to create a colorful chart that compares the sizes of all of the planets within the solar system.

Materials

Microsoft Excel computer spreadsheet application

Procedure

Complete the following steps.

1. Enter the eight planet names shown in Table 9–1 into Column A of your spreadsheet.
2. Enter the "Relative Size Compared to the Earth" numbers for each planet shown in Table 9–1 into Column B of your spreadsheet.
3. Enter the "Relative Size" numbers for each planet shown in Table 9–1 into Column C of your spreadsheet.
4. After all the data have been entered, click into cell A1 of your spreadsheet and select Chart from the Insert Menu. You may also click on the chart wizard icon if available on your toolbar.
5. Select "Bubble Chart with 3-D Visual Effect" in the "Chart Type" window, and click the "Next" button.
6. Click the "Next" button in the "Chart Source Data" window, then fill in the Chart Title in the "Chart Title" box. Click on "Gridlines" and remove check marks from the gridlines box. Click on "Axes" and remove check marks from value X & Y box. Also click "Legend" and remove check marks from "Show Legend" box.
7. In the "Data Labels" window, check X value, and click the "Next" button.
8. Save your chart "As a New Sheet" and click "Finish."
9. Your chart should now show the relative sizes of the planets.
10. If you click once on your chart and click again on a specific planet, you can change its colors. Your chart is now complete.

	A	B	C
		TABLE 9–1	
		Relative Size Compared to Earth	Relative Size
1	Planet		
2	Mercury	0.38	0.38
3	Venus	0.95	0.95
4	Earth	1	1
5	Mars	0.53	0.53
6	Ceres	0.07	0.07
7	Jupiter	11.2	11.2
8	Saturn	9.41	9.41
9	Uranus	4.11	4.11
10	Neptune	3.81	3.81
11	Pluto	0.18	0.18
12	Eris	0.19	0.19

Conclusions

1. Using your chart, state the general relationship between the size of a planet in the solar system and its distance from the Sun.

2. What are the four terrestrial planets, and how do their sizes compare to the gaseous planets?

3. Which is the largest planet in the solar system, and how many times larger is it than the Earth? Show your work.

4. If you were to add the relative size of the Sun to your chart, how would its size compare to all the other planets?

LAB 10
Ellipses and Orbital Motion

Purpose

The purpose of this lab is to have you become familiar with the orbital motion of celestial objects and be able to describe them in the form of an ellipse. In this lab, you will be able to determine the eccentricity of an ellipse and identify how a celestial object's velocity is effected by its orbital position during its orbit.

Materials

ruler
colored pencils
protractor
graph paper

Procedure A

1. Figure 10–1 represents a scale model of the elliptical orbit of the Earth as it revolves around the Sun. Determine the eccentricity of the orbit by using the following formula:

$$\text{eccentricity} = \frac{\text{distance between foci}}{\text{length of the major axis}}$$

Determine the distance between foci for the orbit of the Earth by measuring the distance between the Sun (f_1) and f_2. Record the distance to the nearest tenth of a centimeter in Table 10–1.

TABLE 10–1							
Planet	Length of Major Axis	Distance Between Foci	Eccentricity Calculation	Eccentricity	Angular Velocity near Aphelion	Angular Velocity near Midpoint	Angular Velocity near Perihelion
Earth							
Mercury							

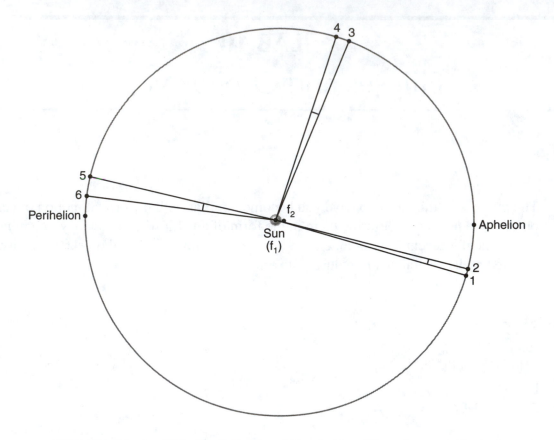

FIGURE 10–1 Elliptical Orbit of Earth (1 million km = 0.7 mm)

2. Next, determine the length of the major axis by measuring the distance from the perihelion point of the orbit through the foci to the aphelion point of the orbit. Record the distance to the nearest tenth of a centimeter in Table 10–1.

3. Now calculate the eccentricity of the Earth's orbit by using the eccentricity formula. Show your calculation and record your answer to the nearest thousandth in Table 10–1.

4. Next, you will calculate the relative orbital velocity of the Earth in three different parts of its orbit around the Sun by determining the angle that it moves through space during a specific time. This is called the angular velocity. Using a protractor and the orbit of the Earth in Figure 10–1, place the vertex of your protractor on the Sun's position (f_1) and determine the angle between lines 1-2 (near aphelion), 3-4 (near midpoint), and 5-6 (near perihelion) on the diagram. Record the angle for each of the three positions to the nearest whole degree in Table 10–1.

5. Repeat procedures 1 through 4 using Figure 10–2, which depicts the elliptical orbit of the planet Mercury around the Sun.

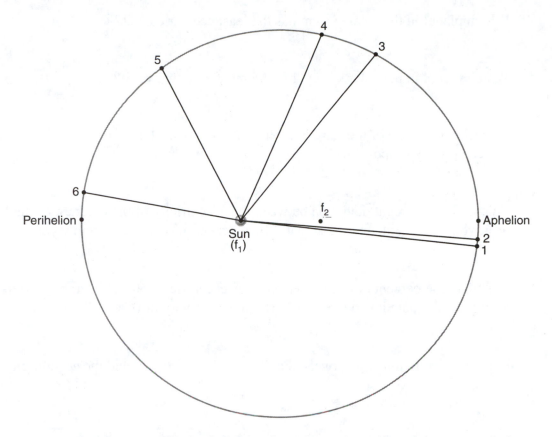

FIGURE 10–2 Elliptical Orbit of Mercury (1 million km = 1.8 mm)

Procedure B

Using your data on the angular velocity of each planet at three points in their orbits, create two bar graphs that show the relationship between the three orbital positions and the angular velocity for both the Earth and Mercury. The three orbital positions are the perihelion, midpoint, and the aphelion. The *x*-axis should be labeled "Orbital Position," and the *y*-axis should be labeled "Relative Angular Velocity." Fit both graphs on one piece of graph paper so that they can be compared.

Conclusions

1. Describe what happens to the shape of an ellipse when its eccentricity increases.

2. What is the name of the object that has an eccentricity of zero?

3. What is the name of the object that has an eccentricity of 1?

4. Which planet in the solar system has the least eccentric orbit?

5. Which planet in the solar system has the most eccentric orbit?

6. Describe the shape of the Earth's orbit. Is it more oval shaped or circular, and how does its eccentricity support this?

7. Calculate the percent deviation between the eccentricity of the diagram of the Earth's orbit you calculated, and the actual orbital eccentricity of 0.017. Show your work.

8. Calculate the percent deviation between the eccentricity of the diagram of Mercury's orbit you calculated, and the actual orbital eccentricity of 0.206. Show your work.

9. Explain the relationship between the distance between foci and the eccentricity of an ellipse.

10. What is the eccentricity of the orbit of the planet Mars, where the distance between foci is 42.4 million kilometers and the length of the major axis is 455.8 million kilometers?

11. Define the terms *aphelion* and *perihelion*. At what times during the year does the Earth reach these two positions?

12. Describe the relationship between the distance of a planet from the Sun and its orbital velocity.

LAB 11
Solar System Math

Purpose

The purpose of this lab is to have you use your math skills to analyze data about the planets in the solar system.

Materials

graph paper or spreadsheet computer application

Procedure

Using the data from Tables 11–1 and 11–2, and the conversion formulas provided, fill in the information on the solar system in Table 11–3.

Using the data from Tables 11–1, 11–2, and 11–3, plot a line graph that shows the relationship between a planet's surface temperature and its distance from the Sun. Your x-axis will be labeled "Distance from the Sun in Millions of Kilometers" and your y-axis will be labeled "Surface Temperature in Degrees Fahrenheit."

TABLE 11–1 Planetary Data				
Name	Gravity (g)	Esc. vel. (km/s)	M.O.V. (km/s)	Surface (K)
Mercury	0.378	4.44	47.87	440
Venus	0.907	10.36	35.02	730
Earth	1.000	11.19	29.79	287
Mars	0.377	5.03	24.13	218
Jupiter	2.364	59.5	13.06	120
Saturn	0.916	35.5	9.66	88
Uranus	0.889	21.3	6.80	59
Neptune	1.125	23.5	5.44	48

KEY:
Gravity   Equatorial surface gravity in g's
Esc. Vel.   Escape velocity in kilometers per second
M.O.V.   Mean orbital velocity in kilometers per second
Surface   Surface temperature in kelvins

TABLE 11–2 Solar System Data Table

Object	Mean Distance from Sun (millions of km)	Period of Revolution	Period of Rotation	Eccentricity of Orbit	Equatorial Diameter (km)	Mass (Earth 5 1)	Density (g/cm³)	Number of Moons
Sun	–	–	27 days	–	1,392,000	333,000.00	1.4	–
Mercury	57.9	88 days	59 days	0.206	4,880	0.553	5.4	0
Venus	108.2	224.7 days	243 days	0.007	12,104	0.815	5.2	0
Earth	149.6	365.26 days	23 hr 56 min 4 sec	0.017	12,756	1.00	5.5	1
Mars	227.9	687 days	24 hr 37 min 23 s	0.093	6,787	0.1074	3.9	2
Jupiter	778.3	11.86 years	9 hr 50 min 30 sec	0.048	142,800	317.896	1.3	39
Saturn	1,427	29.46 years	10 hr 14 min	0.056	120,000	95.185	0.7	18
Uranus	2,869	84.0 years	17 hr 14 min	0.047	51,800	14.537	1.2	21
Neptune	4,496	164.8 years	16 hr	0.009	49,500	17.151	1.7	8
Earth's Moon	149.6 (0.386 from Earth)	27.3 days	27 days 8 hr	0.055	3,476	0.0123	3.3	–

TABLE 11–3 Solar System Data

Planet	Mean Orbital Velocity (miles per hour)	Average Surface Temperature (°F)	Period of Revolution (miles)	Period of Revolution in Sci. Notation

Conclusions

1. Describe the relationship between a planet's mean orbital velocity and its distance from the Sun.

2. Describe the relationship between a planet's distance from the Sun and its period of revolution.

3. Which planet has the greatest orbital velocity?

4. Describe the relationship between a planet's average surface temperature and its distance from the Sun.

5. Which planet is closest to the Sun, and is it the planet with the highest average surface temperature? Why or why not?

6. The Asteroid Belt is located between the orbits of the planets Mars and Jupiter. Some astronomers believe that this is the remains of another planet that was broken apart by the gravitational influence of Mars and Jupiter. If this was once a planet, what do you believe its mean orbital velocity (mph), average temperature (°F), and period of revolution (miles) would have been?

LAB 12
Doppler Shift and the Changing Universe

Purpose

The purpose of this lab is to have you use the spectral analysis of different galaxies in the universe to calculate their relative movement, velocity, and distance from the Earth as determined by Doppler shift. This information can then be used to infer about the present state of the universe. Doppler shift is a change in the wavelength of light emitted from a celestial object that is moving relative to an observer (see Figure 12–1). If the object's motion is toward the observer, the wavelength is reduced, also known as a blue shift. If the object is moving away from the observer, the wavelength is increased, also called a red shift. Astronomers analyze the light emitted from stars and galaxies by using a spectroscope, which reveals the

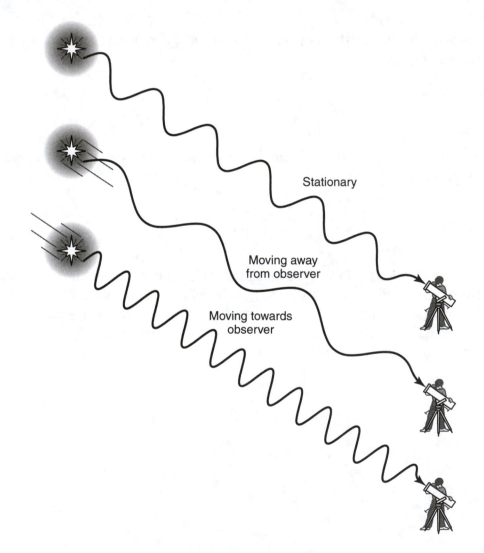

Stationary

Moving away
from observer

Moving towards
observer

FIGURE 12–1   Doppler Shift

object's spectrum. Lines in the spectrum reveal the different types of elements that are present in the object, which occupy a specific location within the spectrum. Astronomers then compare the spectral lines of a star or galaxy with a laboratory standard. This often reveals a shift in the location of the spectral lines caused by Doppler shift. The specific change in location, or shift in the spectral lines, can then be used to determine the relative motion of the object, its velocity, and how far away the object is.

Materials

ruler
colored pencils

Procedure A

1. Use a red colored pencil to draw a line on each galaxy spectrum shown in Figure 12–2, which represents the location of the hydrogen spectral line as listed in Table 12–1. For example, the Virgo A galaxy's hydrogen spectral line should be drawn at the 658.8 nanometer position.

2. Now that you have drawn in all of the galaxies spectral lines for hydrogen as observed through spectral analysis, add the location of the laboratory hydrogen spectral line to act

TABLE 12–1						
Galaxy	Hydrogen Spectral Line Location (nm)	Wavelength Shift	Velocity Calculation	Velocity km/s	Distance Calculation	Distance from Earth (MLy)
Virgo A	658.8					
M65 Spiral Galaxy in Leo	657.8					
The Coma Pinwheel Galaxy	661.3					
Cetus A	658.5					
M58 Spiral Galaxy in Virgo	659.3					
M109 Spiral Galaxy in Ursa Major	658.3					
Andromeda	655.3				—	2.9

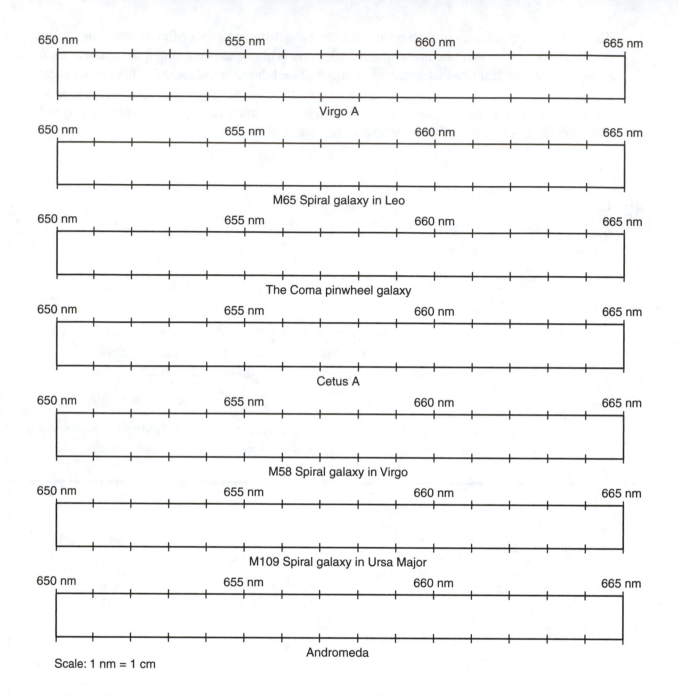

650 nm · 655 nm · 660 nm · 665 nm

Virgo A

M65 Spiral galaxy in Leo

The Coma pinwheel galaxy

Cetus A

M58 Spiral galaxy in Virgo

M109 Spiral galaxy in Ursa Major

Andromeda

Scale: 1 nm = 1 cm

FIGURE 12–2   Absorption Spectrums of Various Galaxies (650–665 nm range)

as a reference point. The spectral line for laboratory hydrogen represents how the spectrum should appear for an object that is not in motion; therefore, any change in the location of the hydrogen line observed in a galaxy or star spectrum represents a Doppler shift. The laboratory standard spectral line for hydrogen is 656 nanometers. Using a green colored pencil, draw in the laboratory hydrogen spectral line in the correct location on each galaxy's spectrum in Figure 12–2.

3. Next, the specific wavelength difference between the laboratory hydrogen and galaxy's hydrogen spectral lines can be used to determine a galaxy's radial velocity. The radial velocity is the speed at which the galaxy is moving either directly toward or away from an observer. Using a metric ruler, determine the difference in wavelength to the nearest tenth, between the laboratory hydrogen and each galaxy's hydrogen spectral line using the equivalent of 1 centimeter = 1 nanometer. If the wavelength of the galaxy's hydrogen is greater than the laboratory hydrogen, record the wavelength difference as a positive number to the nearest tenth of a nanometer in red pencil in the Wavelength Shift column of Table 12–1. If the wavelength of the galaxy's hydrogen is less than the laboratory hydrogen line, record the wavelength difference as a negative number in blue pencil in Table 12–1.

4. The shift in wavelength between the laboratory hydrogen and galaxy's hydrogen spectral lines can now be used to calculate the radial velocity of each galaxy using the following formula:

$$\frac{\text{radial velocity (km/s)}}{300,000 \text{ km/sec}} = \frac{\text{wavelength shift}}{656 \text{ nm}}$$

Enter the wavelength shift value into the formula, and then solve for the radial velocity for each galaxy. Show your calculations and record your answers in Table 12–1.

5. Now that you have determined the velocity for each galaxy, use it to calculate the relative distance from the Earth using Hubble's law. Hubble's law, also known as The Law of Red Shifts, was developed by astronomers Edwin Hubble and Milton Humason in 1931. It states that the distances to galaxies are proportional to their velocities. Use the following Hubble's law formula and the galactic velocities you determined from step 4 to determine the relative distance of each galaxy from the Earth. The *25 km/s/million light years* value in the formula represents the Hubble constant.

$$\text{Distance (millions of light years)} = \frac{\text{radial velocity (km/s)}}{25 \text{ km/s/million light years}}$$

Record the distance for each galaxy to the nearest tenth in the Distance from Earth column in Table 12–1. The distance for the Andromeda galaxy has already been calculated because Hubble's law applies only to receding galaxies.

Procedure B

1. Use your data from Table 12–1 on the velocity and distance for the seven galaxies you analyzed in Procedure A to create a line of best fit graph. The *x*-axis should be labeled "Distance (millions of light years)," and the *y*-axis should be labeled "Velocity (kilometers/second)."

Conclusions

1. Describe how an increase or a decrease in the wavelength of spectral lines from a galaxy or star can be used to infer its motion.

2. When the wavelength of a spectral line emitted from an object increases, which end of the visible light spectrum does it move toward, and what is the object's motion relative to Earth?

3. When the wavelength of a spectral line emitted from an object decreases, which end of the visible light spectrum does it move toward, and what is the object's motion relative to Earth?

4. What did the results of your analysis of the spectral lines of the seven galaxies reveal about their motion relative to Earth?

5. Describe two ways that the Andromeda galaxy differs from the other six galaxies you examined.

6. Do the results of your study of galactic motion suggest that the universe is currently in motion, and if so, is it expanding or contracting, and why?

7. Explain how the results of your study of galactic motion either support or refute the big bang theory.

8. What is the relationship between the distance of a galaxy from the Earth and its velocity?

LAB 13
Density of the Earth's Crust

Purpose

The purpose of this lab is to have you determine the differences in density of the Earth's continental and oceanic crusts. The density differences of these two general forms of the Earth's lithosphere have a major impact on the interactions of the planet's tectonic plates.

Materials

10 samples of oceanic crust (basalt, approximately one to two inches in diameter) for each student group
balances or scales

10 samples of continental crust (granite, approximately one to two inches in diameter) for each student group
250-ml plastic graduated cylinders

Procedure

Complete the following steps.

1. Record the mass and volume of ten different samples of oceanic crust in Table 13–1. Always determine the mass of each sample first, then use a graduated cylinder and the displacement method for determining the volume of each sample. The displacement method involves the filling of a graduated cylinder with a known volume of water. Then, slowly slide your rock sample into the cylinder containing the known volume of water and record the new volume. Subtract the start volume of the water in the cylinder from the volume of the water after you placed the rock sample in the cylinder. The difference between the two volumes is the volume of your rock sample. This method is often used to determine the volume of irregularly shaped objects.
2. Using the data on the mass and volume for each sample of oceanic crust, calculate the density for each sample (density = mass/volume). Record your answers in Table 13–1.
3. Use the ten densities of each sample to determine the average density of oceanic crust. Record your answers in Table 13–1.
4. Record the mass and volume of ten different samples of continental crust in Table 13–2. Always determine the mass of each sample first, then use a graduated cylinder and the displacement method for determining the volume of each sample.
5. Using the data on the mass and volume for each sample of continental crust, calculate the density for each sample (density = mass/volume). Record your answers in Table 13–2.
6. Use the ten densities of each sample to determine the average density of continental crust. Record your answers in Table 13–2.

TABLE 13–1   Oceanic Crust

Sample Number	Mass (grams)	Volume (cubic centimeters)	Density
1			
2			
3			
4			
5			
6			
7			
8			
9			
10			
Average			

TABLE 13–2   Continental Crust

Sample Number	Mass (grams)	Volume (cubic centimeters)	Density
1			
2			
3			
4			
5			
6			
7			
8			
9			
10			
Average			

Conclusions

1. Which type of Earth's crust is more dense? Less dense?

2. Which type of rock makes up the continental crust?

3. Which type of rock makes up the oceanic crust?

4. Describe what would occur if a tectonic plate composed of oceanic crust converged with a tectonic plate composed of continental crust. How is this influenced by their differences in density?

5. Describe the differences in density and rock types of oceanic and continental crust, and explain how these differences influence the interactions of tectonic plates.

LAB 14
The Earth's Interior

Purpose

The purpose of this lab is to have you identify the physical properties that are associated with the different portions of the Earth's interior.

Materials

graph paper or
    spreadsheet
computer application

colored pencils

Procedure

Complete the following steps.

1. Use the data provided in Table 14–1 to construct a line graph showing the relationship between the temperature of the Earth's interior and depth. The x-axis should be labeled "Depth (km)" and the y-axis should be labeled "Temperature of the Earth's Interior (Celsius)."
2. Once all of your data points have been plotted, connect the points with a colored pencil and label the line "Temperature of the Earth's Interior."
3. On the same line chart you just created, add the data points showing the melting point for the Earth's mantle shown in Table 14–1.
4. Once all of the data points have been plotted, connect them with a different color and label the line "Mantle Melting Point Temperature."
5. On the same line chart, add the data points showing the melting point for the Earth's core as shown in Table 14–1.
6. Once all of the data points have been plotted, connect them with a different color and label the line "Core Melting Point Temperature."
7. Using the information from Table 14–2, shade in and label the different portions of the Earth's interior with five different colored pencils.

TABLE 14–1 Earth's Interior Temperature and Depth

Depth (km)	Temperature of the Earth's Interior (°C)	Mantle Melting Point Temperature (°C)	Core Melting Point Temperature (°C)
0	0		
200	1200	1100	
400	2200	2000	
600	2600	2600	
800	2900	3400	
1000	3200	4200	
1200	3400	4800	
1400	3700	5500	
1600	3900	5900	
1800	4100	6200	
2000	4300		
2200	4400		
2400	4600		
2600	4700		
2800	4800		
3000	5000		4400
3200	5200		4700
3400	5300		4900
3600	5400		5000
3800	5600		5200
4000	5700		5400
4200	5800		5600
4400	5900		5700
4600	6000		5800
4800	6200		6000
5000	6300		6300
5200	6400		6500
5400	6500		6700
5600	6600		6800
5800	6700		7000
6000	6800		7000

TABLE 14–2	Earth's Interior Zones
Zone	Depth (km)
Lithosphere	0–120
Asthenosphere	120–600
Mantle	600–2900
Outer Core	2900–5200
Inner Core	5200–6400

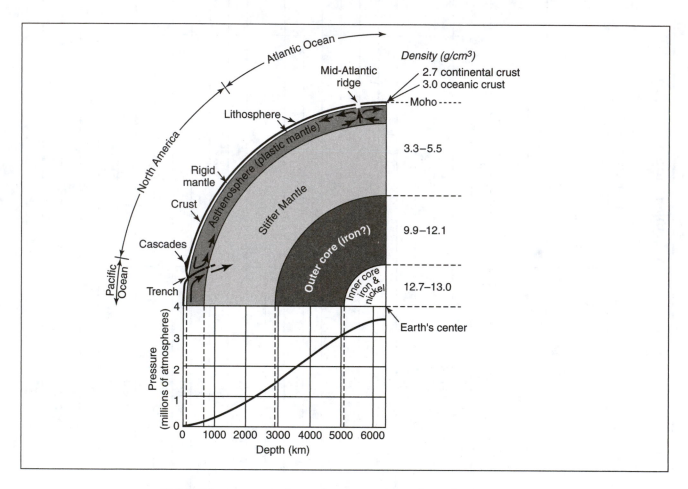

FIGURE 14–1   Inferred Properties of Earth's Interior

Conclusions

Use the line chart you have just created along with the Inferred Properties of the Earth's Interior (Figure 14–1) to answer the following questions.

1. What is the average density of the Earth's crust?

2. Describe the temperature range within the Earth's outer crust.

46

3. What is the approximate depth of the Earth's outer crust?

4. What is the range of temperatures that exists within the asthenosphere?

5. Does the temperature of the rock within the asthenosphere meet the melting point?

6. What is the state of matter of the rock within the asthenosphere?

7. What is the range of depth of the asthenosphere?

8. What is another name for the asthenosphere?

9. What is the range of density within the stiffer mantle?

10. What is the range of temperatures within the stiffer mantle?

11. Is the temperature of the stiffer mantle above or below its melting point?

12. What is the state of matter of the rock within the stiffer mantle?

13. At what range of depth does the stiffer mantle exist?

14. What is the range of pressure within the stiffer mantle?

15. What is the outer core believed to be composed of?

16. What is the range of density within the outer core?

17. What is the temperature range within the outer core?

18. What is the pressure range within the outer core?

19. Is the temperature of the outer core higher or lower than its melting point?

20. What is the state of matter within the outer core?

21. What is the range of depth of the outer core?

22. What is the range of density within the inner core?

23. What is the range of temperatures within the inner core?

24. What is the inner core believed to be composed of?

25. What is the range of depth for the inner core?

26. Is the temperature of the inner core lower or higher than its melting point?

27. What state of matter is the inner core?

LAB 15
Sea Floor Spreading

Purpose

The purpose of this lab is to have you use the age and positions of rock formations associated with the Mid-Atlantic Ridge to support the hypothesis of sea floor spreading. The sea floor spreading hypothesis presents the idea that new crust is continually being formed at the center of mid-ocean ridges, which is causing the sea floor to spread apart from the ridge centers. This is believed to be the mechanism that causes the Earth's tectonic plates to move.

Materials

metric ruler   colored pencils

Procedure

Complete the following steps.

1. Using the map in Figure 15–1, color in the various rock formations associated with both the Mid-Atlantic and East Pacific Ridge systems. Create a color key that shows the color associated with each specific rock age.
2. After you have created a map showing the relationship between the age of rock and location to the ridge system center, calculate an average spreading rate associated with each ridge. To calculate the average spreading rate for the Mid-Atlantic Ridge, use the following procedure. Show all work.
    a. Measure the distance in centimeters from the Mid-Atlantic Ridge to the interface between rock sequences 4 and 5 in three different places on your map. Record your answers below.

    Measurement 1 =

    Measurement 2 =

    Measurement 3 =

    b. Determine the average distance of the three measurements to the nearest tenth of a centimeter. Show your work and record your average distance to the nearest tenth of a centimeter.

    Average distance (cm) =

c. Convert your average measurement in centimeters to kilometers by using the map scale of 1 centimeter = 500 kilometers. This is done by multiplying the average distance in centimeters by 500 kilometers. Record your answer to the nearest tenth of a kilometer.

Average distance (km) =

d. Knowing that the age of the interface between rock sequences 4 and 5 are approximately 118 million years old, calculate how far the rocks have moved in kilometers per year. This is done by dividing your average distance in kilometers by 118 million years. Record your answer.

Distance rock moved (km per year) =

e. Now, determine the distance the rock moved from the ridge center in centimeters by multiplying the distance the rock moved in kilometers per year by 100,000 (100,000 centimeters per kilometer). Record your answer.

Distance rock moved (cm per year) =

f. Determine the rate of sea floor spreading associated with the Mid-Atlantic Ridge by multiplying the distance the rock moved in centimeters per year by 2. This determines the rate of sea floor spreading because new rock is created on both sides of the ridge and is pushed outward. Record your answer.

Rate of sea floor spreading for the Mid-Atlantic Ridge (cm per year) =

g. Finally, convert the above spreading rate into inches per year by dividing it by 2.5 (2.5 cm = 1 inch). Record your answer.

Rate of sea floor spreading for the Mid-Atlantic Ridge (inches per year) =

3. To calculate the average spreading rate for the East Pacific Ridge, use the following procedure. Show all work.
   a. Measure the distance in centimeters from the East Pacific Ridge to the interface between rock sequences 3 and 4 in three different places on your map. Record your answers.

Measurement 1 =

Measurement 2 =

Measurement 3 =

b. Determine the average distance of the three measurements to the nearest tenth of a centimeter. Show your work and record your average distance to the nearest tenth of a centimeter.

Average distance (cm) =

c. Convert your average measurement in centimeters to kilometers by using the map scale of 1 centimeter = 500 kilometers. This is done by multiplying the average distance in centimeters by 500 kilometers. Record your answer to the nearest tenth of a kilometer.

Average distance (km) =

d. Knowing that the age of the interface between rock sequences 3 and 4 are approximately 56 million years old, calculate how far the rocks have moved in kilometers per year. This is done by dividing your average distance in kilometers by 56 million years. Record your answer.

Distance rock moved (km per year) =

e. Now, determine the distance the rock moved from the ridge center in centimeters by multiplying the distance the rock moved in kilometers per year by 100,000 (100,000 centimeters per kilometer). Record your answer.

Distance rock moved (cm per year) =

f. Determine the rate of sea floor spreading associated with the East Pacific Ridge by multiplying the distance the rock moved in centimeters per year by 2. This determines the rate of sea floor spreading because new rock is created on both sides of the ridge and is pushed outward. Record your answer.

Rate of sea floor spreading for the East Pacific Ridge (cm per year) =

g. Finally, convert the above spreading rate into inches per year by dividing it by 2.5 (2.5 cm = 1 inch). Record your answer.

Rate of sea floor spreading for the East Pacific Ridge (inches per year) =

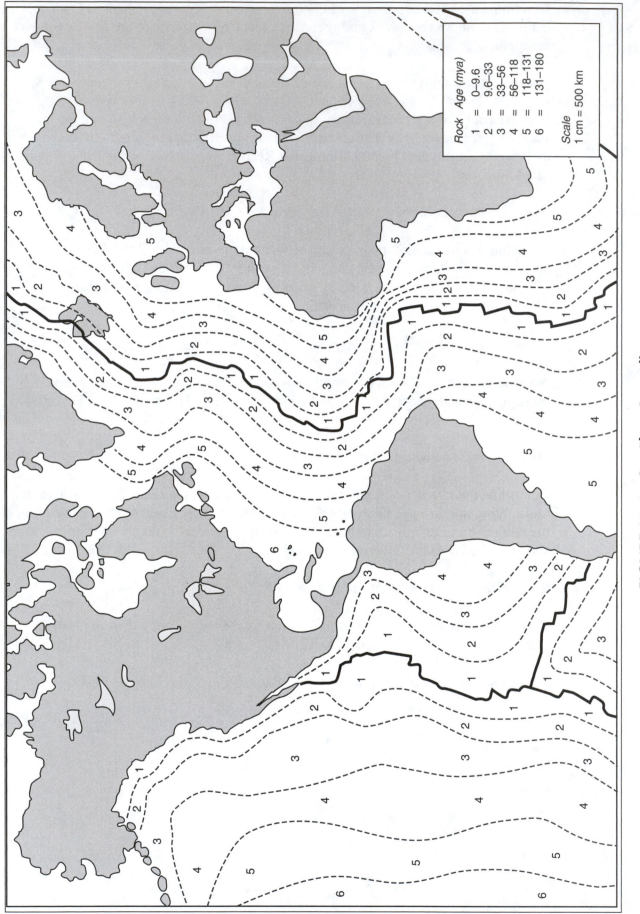

FIGURE 15–1   Sea Floor Spreading

Rock   Age (mya)
1 = 0–9.6
2 = 9.6–33
3 = 33–56
4 = 56–118
5 = 118–131
6 = 131–180

Scale
1 cm = 500 km

Conclusions

1. Describe how the mapping of rock ages on the ocean floor helps support the theory of continental drift and sea floor spreading.

2. Briefly explain how the rate of sea floor spreading can be determined in both the Atlantic and Pacific Oceans.

3. Are the average rates of sea floor spreading the same in both the Atlantic and Pacific Oceans, and if not, how do they compare?

4. Explain how the discovery of sea floor spreading has helped the theory of continental drift.

LAB 16
Tectonic Plate Boundaries

Purpose

The purpose of this lab is to have you identify the three ways in which tectonic plates can interact, and how these interactions explain the geologic processes of mountain building, volcanoes, deep ocean trenches, mid-ocean ridges, rift valleys, earthquakes, and sea floor spreading.

Materials

colored pencils   map of the tectonic plates

Procedure

Complete the following steps.

1. Using the map in Figure 16–1 that shows the tectonic plates from your textbook, shade in the three different plate boundaries and label their names on the map provided. Use a different colored pencil for each different type of plate boundary. Create a key that shows the color that represents each plate boundary.
2. Choose another colored pencil to add the arrows of movement associated with each plate boundary.
3. Label all of the names of the major tectonic plates in another color on your map.
4. Use your tectonic plate map to identify the different type of plate boundaries and the specific tectonic plates that form each geologic region shown in Table 16–1.

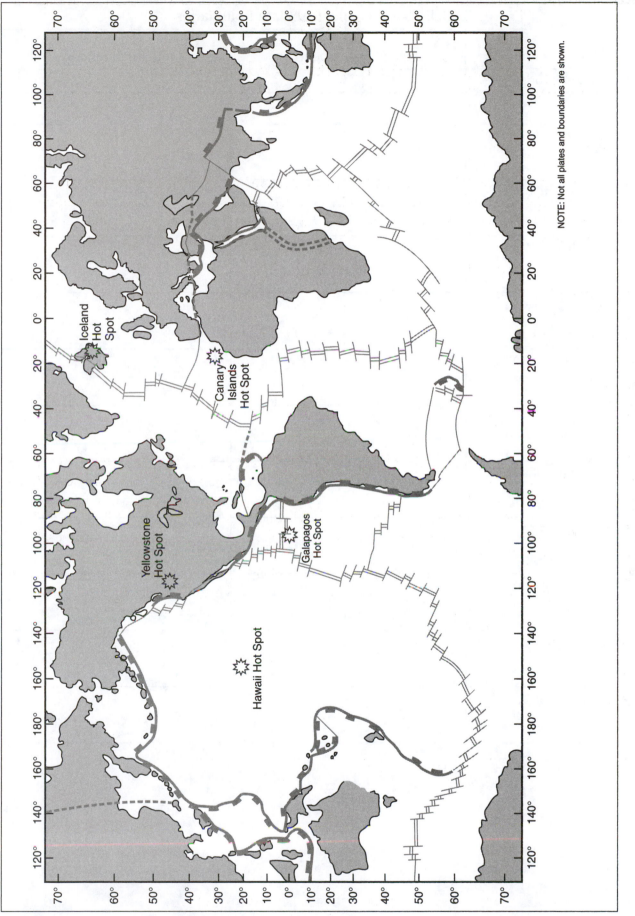

Iceland
Hot
Spot

Canary
Islands
Hot Spot

Yellowstone
Hot Spot

Galapagos
Hot Spot

Hawaii Hot Spot

NOTE: Not all plates and boundaries are shown.

FIGURE 16–1   Tectonic Plates

TABLE 16–1 Geological Plate Boundaries		
Geologic Region	Type of Plate Boundary	Tectonic Plates Associated with the Boundary
Mid-Atlantic Ridge		
Tonga Trench		
East African Rift		
Peru-Chili Trench		
East Pacific Ridge		
Mariana Trench		
San Andreas Fault		
Himalaya Mountains		
Andes Mountains		
Cascade Mountains		
Aleutian Trench		
Mount St. Helens		

Conclusions

1. What are the three different types of tectonic plate boundaries?

2. Which geologic features on the Earth are associated with divergent plate boundaries?

3. Describe the three ways in which the Earth's crust can interact at convergent plate boundaries.

4. Explain the tectonic process known as subduction.

5. Which geologic features are associated with convergent plate boundaries?

6. Describe the interactions of tectonic plates around a transform fault boundary.

7. What geologic feature is associated with a transform fault boundary in the United States?

LAB 17
Locations of Mountains, Earthquakes, and Volcanoes

Purpose

The purpose of this lab is to identify the locations of earthquakes, volcanoes, and mountain ranges on the Earth, and their relationships to locations of tectonic plate boundaries.

Materials

world maps or globes   colored pencils

Procedure

Complete the following steps.

1. Using the data in Table 17–1, plot the locations of earthquake epicenters around the world with an "E" using a specific colored pencil on the map in Figure 17–1.
2. Using the data in Table 17–1, plot the locations of volcanoes with a "V" using a specific colored pencil on the map.
3. On your map, label the following mountain ranges using a third colored pencil: the Andes, Himalayas, and Cascade Mountains.
4. Using the information from Table 17–1, plot the location and label the following volcanoes on your map: Mount St. Helens, Mount Vesuvius, Mauna Loa, and Krakatau.

TABLE 17–1 Latitude/Longitude Location of Earthquakes

Latitude	Longitude	Latitude	Longitude	Latitude	Longitude
43 S	97 E	50 S	170 W	46 S	30 E
50 S	120 E	47 S	160 W	45 S	44 E
52 S	130 E	46 S	145 W	40 S	52 E
53 S	150 E	41 S	137 W	35 S	65 E
54 S	157 E	38 S	131 W	30 S	72 E
55 S	170 E	32 S	123 W	22 N	75 E
54 S	178 E	25 S	121 W	22 N	80 E
52 S	30 W	7 S	110 W	20 N	85 E
49 N	30 W	2 N	110 W	18 N	100 E
45 N	32 W	10 N	110 W	10 N	97 E
37 N	35 W	20 N	110 W	0	98 E
33 N	37 W	20 N	105 W	5 S	100 E
23 N	39 W	23 N	110 W	10 S	110 E
15 N	53 W	30 N	118 W	15 S	112 E
7 N	30 W	52 S	75 W	16 S	125 E
0	25 W	48 S	75 E	13 S	130 E
7 S	20 W	43 S	74 E	10 S	130 E
15 S	15 W	31 S	70 E	5 N	130 E
22 S	15 W	25 S	70 E	0	42 E
30 S	15 W	22 S	70 E	10 N	85 W
37 S	13 W	15 S	75 E	13 N	90 W
44 S	12 W	25 S	75 E	15 N	97 W
50 S	3 W	30 S	80 E		
50 S	8 E	37 S	90 E		
82 N	14 E	10 N	130 E		
79 N	7 E	10 N	145 E		
76 N	4 E	15 N	145 E		
72 N	0	20 N	127 E		
70 N	7 W	25 N	128 E		
67 N	14 W	30 N	135 E		
65 N	18 W	0	150 E		
63 N	22 W	5 S	160 E		
58 N	28 W	10 S	170 E		
59 N	178 E	15 S	178 E		
56 N	170 E	35 S	118 E		
50 N	165 E	40 N	123 E		
48 N	157 E	45 N	125 E		
46 N	150 E	48 N	127 W		
37 N	148 E	55 N	137 W		
35 N	142 E	58 N	142 W		
30 N	140 E	55 N	150 W		
52 S	25 W	51 N	155 W		
50 S	20 W	50 N	160 W		
48 S	11 W	50 N	170 W		
7 N	65 E	7 S	82 W		
1 S	70 E	0	80 W		
11 S	72 E	7 N	77 W		
20 S	75 E	49 S	20 E		

VOLCANOES

Latitude	Longitude
65 N	23 W
52 N	30 W
41 N	33 W
30 N	37 W
10 N	53 W
13 S	18 W
38 S	15 W
45 N	125 W
5 S	80 W
15 S	75 W
35 N	140 E
5 S	155 E
7 N	128 E

Mount St. Helens: 46 N–122 W
Mount Vesuvius: 41 N–14 E
Mauna Loa: 19 N–156 W
Krakatau: 6 S–105 E

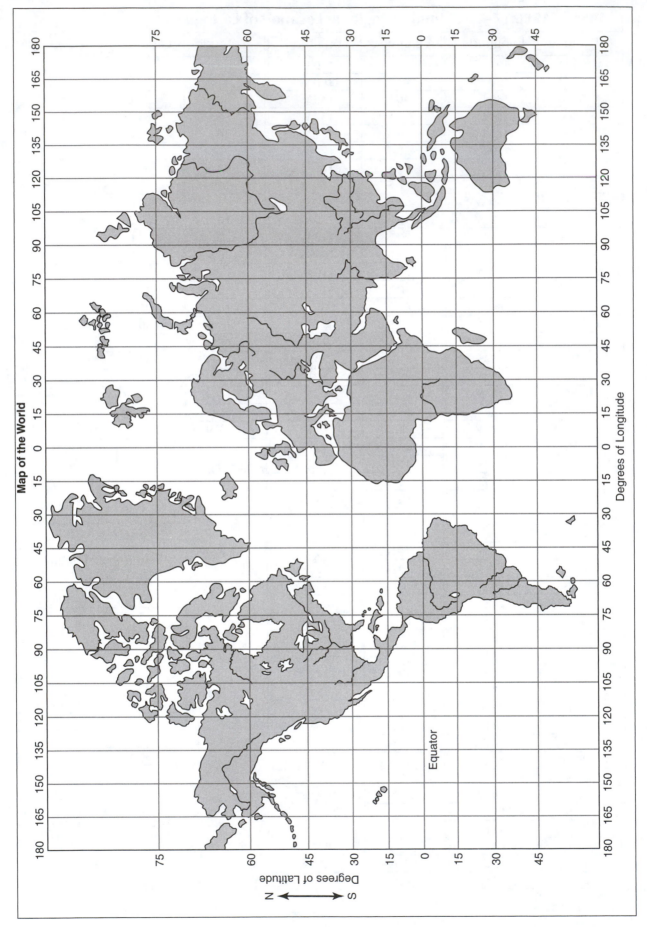

Map of the World

Degrees of Longitude

Degrees of Latitude

N ← → S

Equator

FIGURE 17–1   Map of the World

60

Conclusions

1. Describe the pattern of earthquake, volcano, and mountain range locations on the Earth.

2. Explain how the locations of earthquakes, volcanoes, and mountain ranges are related to tectonic plate boundaries.

3. Which parts of the United States are most likely to experience earthquakes and volcanic activity?

4. Which "hot spot" volcano on your map is not located near a plate boundary?

LAB 18
Speed of Seismic Waves and Earthquake Epicenter Location

Purpose

The purpose of this lab is for you to identify the different velocities at which seismic waves generated by an earthquake travel through the Earth. You will then use this information to calculate the approximate distance to an earthquake's epicenter.

Materials

graph paper or
  spreadsheet computer application

drawing compass
colored pencils

Procedure

Use the data on the distance and travel time for primary and secondary seismic waves shown in Table 18–1 to construct a line graph. The x-axis will be labeled "Distance (km)" and the y-axis will be labeled "Time (minutes)." Use a different colored pencil for each seismic wave and label each line.

Next, plot another line graph that shows the relationship between the difference in arrival times for both the primary and secondary seismic waves, and the distance from the earthquake epicenter shown in Table 18–2. The x-axis will be labeled "Distance to Epicenter (km)" and the y-axis will be labeled "Difference between P- and S-Wave Arrival Time (minutes)." Use both of your line graphs to answer the conclusion questions.

TABLE 18–1 Primary and Secondary Seismic Waves

Distance to Epicenter (km)	P-Wave Travel Time (minutes)	S-Wave Travel Time (minutes)
0	0	0
200	0.5	1
400	1	1.67
600	1.33	2.67
800	1.67	3.33
1000	2.33	4
1200	2.67	4.67
1400	3	5.33
1600	3.33	6
1800	3.67	6.67
2000	4	7.33
2200	4.33	8
2400	4.67	8.5
2600	5	9
2800	5.33	9.67
3000	5.67	10.17
3200	6	10.67
3400	6.2	11.17
3600	6.5	11.67
3800	6.83	12.17
4000	7	12.67
4200	7.33	13.17
4400	7.67	13.67
4600	7.83	14
4800	8	14.5
5000	8.33	15
5200	8.5	15.33
5400	8.67	15.83
5600	9	16.17
5800	9.17	16.67
6000	9.33	17
6200	9.67	17.33
6400	9.83	17.83
6600	10	18.17
6800	10.33	18.5
7000	10.5	19
7200	10.67	19.33
7400	10.83	19.67
7600	11	20
7800	11.17	20.33
8000	11.33	20.67
8200	11.5	21
8400	11.67	21.33
8600	11.83	21.67
8800	12	22
9000	12.17	22.33
9200	12.33	22.5
9400	12.5	22.67
9600	12.67	23
9800	12.83	23.17
10000	13	23.33

TABLE 18–2   Seismic Waves from Earthquake Epicenter

Approximate Distance to Epicenter (km)	Difference Between P-and S-Wave Arrival Time (minutes)
0	0
800	1
1200	2
1800	3
2800	4
3200	5
4200	6
5400	7
6400	8
7600	9
8800	10

Conclusions

1. Using your graph on P- and S-wave travel times, approximately how long will it take an S-wave to travel 4,000 kilometers?

2. Using your graph on P- and S-wave travel times, approximately how long will it take a P wave to travel 4,000 kilometers?

3. Using your graph on P- and S-wave travel times, approximately how far will a P-wave travel in 3 minutes?

4. Using your graph on P- and S-wave travel times, approximately how far will an S-wave travel in 3 minutes?

5. Based on the data you plotted on your graphs, which seismic wave travels at a greater velocity?

6. How far away from the epicenter of an earthquake are you if the difference between the arrival time of P- and S-waves is 9 minutes?

7. If a P-wave arrives at your location at 3:25 P.M. and the S-wave arrives at 3.32 P.M., how far away is the earthquake epicenter?

8. If a P-wave arrives at your location at 11:15 A.M. and the S-wave arrives at 11:17 A.M., how far away is the earthquake epicenter?

9. Using your graph on the difference between P-and S-wave arrival times, the data below, and help from your instructor, locate the approximate location of an earthquake's epicenter location on the map in Figure 18–1. Use a drawing compass to draw the radius distance from each seismic station where the earthquake occurred. Where the three circles converge is the approximate location of the earthquake epicenter.

Seattle, Washington: P-wave arrival time = 11:18:00 A.M., S-wave arrival time = 11:20:20 A.M.

Miami, Florida: P-wave arrival time = 2:21:20 P.M., S-wave arrival time = 2:26:20 P.M.

Portland, Maine: P-wave arrival time = 2:22:00 P.M., S-wave arrival time = 2:27:40 P.M.

10. Using the data from question 9, determine at what time the earthquake occurred on the west coast (P-wave arrival time – P-wave travel time).

FIGURE 18–1   Earthquake Epicenter Location Map

LAB 19
Mineral Identification

Purpose

The purpose of this lab is to have you identify 21 rock-forming minerals by their unique physical and chemical characteristics. These 21 minerals are important natural resources, and together they form many common rocks found in the Earth's crust.

Materials

magnifying glasses
glass plates
magnets
enough of following mineral samples for each student or student group:

magnetite	pyrite	potassium feldspar (orthoclase)
hematite	talc	streak plates
sulfur	gypsum	steel files
muscovite mica	halite	dilute hydrochloric acid
biotite mica	calcite	graphite
dolomite	flourite	galena
pyroxene (augite)	amphibole (hornblende)	garnet (almandine)
plagioclase feldspar	quartz	olivine

Procedure

SAFETY CONCERNS
THE USE OF HYDROCHLORIC ACID TO IDENTIFY THE MINERAL CALCITE SHOULD BE CONTROLLED BY THE INSTRUCTOR. SAFETY GLASSES SHOULD BE USED WHEN PERFORMING THE ACID TEST! ALSO, MAKE SURE ALL STUDENTS WASH THEIR HANDS THOROUGHLY WITH SOAP AND WATER AFTER HANDLING ALL MINERALS!

Minerals can be identified by determining their unique physical properties. These include luster, hardness, cleavage, color, streak, and other special properties. Use the relative scale of hardness in Table 19–1 and the properties of common minerals in Table 19–2 to help determine each mineral sample's hardness. Then fill in the information for each unknown mineral sample on Table 19–3.

TABLE 19–1   Mohs Scale of Mineral Hardness

Mineral	Hardness	Relative Hardness
Graphite	0.7	Can be scratched by fingernail
Talc	1	
Gypsum	2	
Calcite	3	Can be scratched by copper penny
Flourite	4	Can be scratched by steel
Apatite	5	
Orthoclase	6	Can be scratched by glass
Quartz	7	
Topaz	8	Can be scratched by quartz
Corundum	9	Can be scratched by topaz
Diamond	10	

TABLE 19–2 Properties of Common Minerals

LUSTER	HARD-NESS	CLEAVAGE	FRACTURE	COMMON COLORS	DISTINGUISHING CHARACTERISTICS	USE(S)	MINERAL NAME	COMPOSITION*
Nonmetallic luster	1	★		White to green	Greasy feel	Talcum powder, soapstone	Talc	$Mg_3Si_4O_{10}(OH)_2$
	2		★	Yellow to amber	Easily melted, may smell	Vulcanize rubber, sulfuric acid	Sulfur	S
	2	★		White to pink or gray	Easily scratched by fingernail	Plaster of paris and drywall	Gypsum (Selenite)	$CaSO_4 \cdot 2H_2O$
	2–2.5	★		Colorless to yellow	Flexible in thin sheets	Electrical insulator	Muscovite Mica	$KAl_3Si_3O_{10}(OH_2)$
	2.5	★		Colorless to white	Cubic cleavage, salty taste	Food additive, melts ice	Halite	$NaCl$
	2.5–3	★		Black to dark brown	Flexible in thin sheets	Electrical insulator	Biotite Mica	$K(Mg,Fe)_3$ $AlSi_3O_{10}(OH)_2$
	3	★		Colorless or variable	Bubbles with acid	Cement, polarizing prisms	Calcite	$CaCO_3$
	3.5	★		Colorless or variable	Bubbles with acid when powdered	Source of magnesium	Dolomite	$CaMg(CO_3)_2$
	4	★		Colorless or variable	Cleaves in 4 directions	Hydrofluoric acid	Flourite	CaF_2
	5–6	★		Black to dark green	Cleaves in 2 directions at 90°	Mineral collections	Pyroxene (commonly Augite)	$(Ca,Na)(Mg,Fe,Al)$ $(Si,Al)_2O_6$
	5–5	★		Black to dark green	Cleaves at 56° and 124°	Mineral collections	Amphiboles (commonly Hornblende)	$Ca,Na(Mg,Fe)4(Al,Fe,Ti)3$ $Si_6O_{22}(O,OH)_2$
	6	★		White to pink	Cleaves in 2 directions at 90°	Ceramics and glass	Potassium Feldspar (Orthoclase)	$KAlSi_3O_8$
	6	★		White to gray	Cleaves in 2 directions, striations visible	Ceramics and glass	Potassium Feldspar (Na-Ca Feldspar)	$(Na,Ca)AlSi_3O_8$
	6.5		★	Green to gray or brown	Commonly light green and granular	Furnace bricks and jewelry	Olivine	$(Fe,Mg)_2SiO_4$
	7		★	Colorless or variable	Glassy luster, may form hexagonal crystals	Glass, jewelry, and electronics	Quartz	SiO_4
	7		★	Dark red to green	Glassy luster, often seen as red grains in NYS metamorphic rocks	Jewelry and abrasives	Garnet (commonly Almandine)	$Fe_3Al_2Si_3O_{12}$
Either	1–6.5		★	Metallic silver or earthy red	Red-brown streak	Ore or iron	Hematite	Fe_2O_3
Metallic luster	1–2	★		Silver to gray	Black streak, greasy feel	Pencil lead, lubricants	Graphite	C
	2.5	★		Metallic silver	Very dense (7.6 g/cm3), gray-black streak	Ore of lead	Galena	PbS
	5.5–6.5		★	Black to silver	Attracted by magnet black streak	Ore of iron	Magnetite	Fe_3O_4
	6.5		★	Brassy yellow	Green-black streak, cubic crystals	Ore of sulfur	Pyrite	FeS_2

★ = Dominant form of Breakage

Al = aluminum	Cl = chlorine	H = hydrogen	Na = sodium	S = sulfur
C = carbon	F = fluorine	K = potassium	O = oxygen	Si = silicon
Ca = calcium	Fe = iron	Mg = magnesium	Pb = lead	Ti = titanium

TABLE 19–3   Mineral Properties

Sample	Luster	Hardness	Cleavage	Color	Streak	Special Properties	Name
1							
2							
3							
4							
5							
6							
7							
8							
9							
10							
11							
12							
13							
14							
15							
16							
17							
18							
19							
20							
21							

Conclusions

1. Explain why minerals are not usually identified by their color alone.

2. Which minerals have a metallic luster?

3. Which mineral properties are most useful for identification?

4. What are two specific ways by which you can tell the difference between calcite and halite?

5. What mineral could be used to scratch a piece of potassium feldspar?

LAB 20
Igneous Rock Identification

Purpose

The purpose of this lab is to have you identify common igneous rocks on the basis of their texture, color, density, and mineral composition. You will also be able to identify the environment of formation for each igneous rock sample.

Materials

magnifying glasses
enough of the following igneous rock samples for each student or student group:

basaltic glass	gabbro	pegmatite
pumice	vesicular basalt	diorite
scoria	rhyolite	granite
andesite	basalt	

Procedure

Using your textbook, fill in the missing information on rock names, texture, environment of formation, and mineral composition on the Scheme for Igneous Rock Identification chart in Figure 20–1.

Once your instructor has checked your completed chart for accuracy, use it to identify the igneous rock samples provided. Fill in the appropriate information for each rock in Table 20–1 in order to correctly identify it.

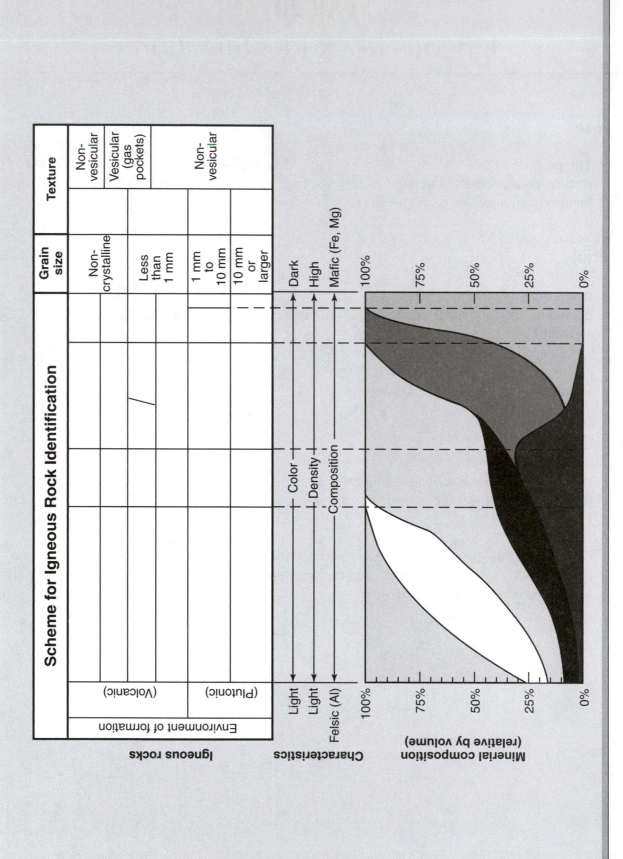

FIGURE 20–1   Scheme for Igneous Rock Identification

TABLE 20–1   Igneous Rock Identification

Rock Name	Texture	Grain Size	Environment of Formation	Color	Density	Mineral Composition	Mafic/ Felsic
1							
2							
3							
4							
5							
6							
7							
8							
9							
10							
11							
12							
13							
14							
15							
16							
17							

Conclusions

1. Explain what is meant by a rock sample's texture.

2. What is the environment of formation for igneous rocks with coarse to very coarse texture?

3. What is the environment of formation for igneous rocks with fine or glassy texture?

4. Explain the difference between lava and magma.

5. Describe the relationship between cooling time and texture for igneous rocks.

6. Describe the specific mineral composition, density, and shade of color of felsic rocks.

7. Describe the specific mineral composition, density, and shade of color of mafic rocks.

8. Which fine-textured igneous rock contains the minerals pyroxene and olivine?

9. What does the term vesicular mean when used to describe the texture of an igneous rock?

10. Which coarse-textured igneous rock contains the minerals quartz and potassium feldspar?

11. Describe the main physical properties that are used to identify igneous rocks.

LAB 21
Sedimentary Rock Identification

Purpose

The purpose of this lab is to have you identify common sedimentary rocks on the basis of their texture, grain size, and composition. You will also be able to identify the processes of formation for each sedimentary rock sample.

Materials

magnifying glasses
enough of the following sedimentary rock samples for each student or student group:

siltstone	shale	sandstone
rock salt	rock gypsum	dilute hydrochloric acid
dolostone	chemical limestone	conglomerate
fossil limestone	bituminous coal	breccia

SAFETY CONCERN
THE USE OF HYDROCHLORIC ACID TO IDENTIFY LIMESTONE SHOULD BE CONTROLLED BY THE INSTRUCTOR. SAFETY GLASSES SHOULD BE USED WHEN PERFORMING THE ACID TEST!

Procedure

Using your textbook, fill in the missing information on texture, grain size, mineral composition, and rock name on the Scheme for Sedimentary Rock Identification chart in Figure 21–1.

Once your instructor has checked your completed chart for accuracy, use it to identify the sedimentary rock samples provided. Fill in the appropriate information for each rock in Table 21–1 in order to identify it. Recall that clastic sedimentary rocks are formed from the processes of cementation and compaction, and crystalline sedimentary rock forms from the processes of evaporation and precipitation.

CHEMICALLY AND/OR ORGANICALLY FORMED SEDIMENTARY ROCKS

Texture	Grain Size	Composition	Comments	Rock Name	Map & Symbol
			Crystals from chemical precipitates and evaporites		
			Cemented shell fragments or precipitates of biologic origin		
			Plant remains		

INORGANIC LAND-DERIVED SEDIMENTARY ROCKS

Texture	Grain Size	Composition	Comments	Rock Name	Map & Symbol
(fragmental)					

FIGURE 21–1   Scheme for Sedimentary Rock Identification

TABLE 21–1   Sedimentary Rock Identification

Texture	Composition	Process of Formation	Method of Lithification	Rock Name

76

Conclusions

1. What are the three types of texture used to identify sedimentary rocks?

2. Describe the differences between clay, silt, and sand sediments.

3. Explain how you can distinguish between sedimentary rocks formed from evaporation and/or precipitation, and those formed from compaction and/or cementation.

4. If one geographic area contains a layer of sandstone while another area contains a layer of shale, what can you infer about the depth of water in which both of these sedimentary rock layers formed?

5. How can you distinguish chemical limestone and dolostone?

6. Explain why you are most likely to only find fossils in sedimentary rocks.

7. Describe the process by which sedimentary rocks are identified.

LAB 22
Metamorphic Rock Identification

Purpose

The purpose of this lab is to have you identify common metamorphic rocks on the basis of their texture, grain size, and degree of metamorphism. You will also be able to identify the parent rock from which each metamorphic rock formed.

Materials

magnifying glasses
enough of the following metamorphic rock samples for each student or student group:

schist	gneiss	dilute hydrochloric acid
quartzite	marble	slate
metaconglomerate	anthracite coal	phyllite

SAFETY CONCERN
THE USE OF HYDROCHLORIC ACID TO IDENTIFY MARBLE SHOULD BE
CONTROLLED BY THE INSTRUCTOR. SAFETY GLASSES SHOULD BE
USED WHEN PERFORMING THE ACID TEST!

Procedure

Using your textbook, fill in the missing information on texture, type of metamorphism, and rock name on the Scheme for Metamorphic Rock Identification chart in Figure 22–1.

Once your instructor has checked your completed chart for accuracy, use it to identify the metamorphic rock samples provided. Fill in the appropriate information for each rock in Table 22–1 in order to identify it.

Texture	Grain Size	Composition	Type of Metamorphism	Comments	Rock Name	Map & Symbol
	Fine	MICA		Low-grade metamorphism of shale		
	Fine to medium	MICA, QUARTZ, FELDSPAR, AMPHIBOLE, GARNET	↓	Foliation surfaces shiny from microscopic mica crystals		
	Fine to medium	MICA, QUARTZ, FELDSPAR, AMPHIBOLE, GARNET, PYROXENE	↓	Platy mica crystals visible from metamorphism of clay or feldspar		
	Medium to coarse	MICA, QUARTZ, FELDSPAR, AMPHIBOLE, GARNET, PYROXENE		High-grade metamorphism; some mica changed to feldspar; segregated by mineral type into bands		
	Fine	Variable		Various rocks changed by heat from nearby magna/lava		
	Fine to coarse	Quartz		Metamorphism of quartz sandstone		
	Fine to coarse	Calcite and/or dolomite		Metamorphism of limestone or dolostone		
	Coarse	Various minerals in particles and matrix		Pebbles may be distorted or stretched		

FIGURE 22–1 Scheme for Metamorphic Rock Identification

TABLE 22–1 Metamorphic Rock Identification

Rock Name	Texture	Grain Size	Type of Metamorphism	Parent Rock

Conclusions

1. Describe the three different types of textures used to identify metamorphic rocks.

2. List all of the foliated metamorphic rocks in order of arrangement as heat, pressure, and depth increase.

3. Explain why most metamorphic rocks are harder and denser than their parent rocks.

4. Describe how you could identify marble from other metamorphic rocks with crystalline texture.

5. If you could find fossils in metamorphic rocks, how would they look and why?

6. Explain why metamorphic rocks are usually found in mountainous regions.

7. Describe the characteristics used to identify metamorphic rocks.

LAB 23
The Rock Cycle

Purpose

The purpose of this lab is to illustrate the various changes that rocks and rock material can go through over time on the Earth. You will also identify the natural processes and pathways that together make up the rock cycle.

Materials

enough of the following igneous, sedimentary, and metamorphic rock samples for each student or student group:

two shale samples	chemical limestone	basalt
bituminous coal	quartzite	two granite samples
marble	gneiss	sandstone
anthracite coal	slate	schist

Procedure

Using your textbook, fill in the missing pathways and processes on the incomplete Rock Cycle diagram in Figure 23–1.

Next, with the help of your instructor, identify all of the rock samples that you have to work with. Once you have correctly identified all of the rocks, ask your instructor to check your completed rock cycle diagram for accuracy. Then use it to help you arrange the various rock samples into groups that illustrate logical sequences that rocks can go through in the rock cycle. An example of a logical sequence that illustrates a portion of the rock cycle would include granite weathering into pebbles and sand, that can eventually form into a conglomerate. As you determine a correct sequence, fill in the names of each rock that forms a logical grouping in the blank flow charts in Figure 23–2. To help you get started, pick out a metamorphic rock sample and determine the parent rock that formed it. Notice that out of the five correct groups, only two contain three sequences while the rest contain two sequences.

After you have arranged all your rock samples into groups, have your instructor check to see if they are correct. Once they are checked for accuracy, use the rock cycle diagram to help you write out in complete sentences, on a separate piece of paper, the specific processes that each grouping goes through to change from one rock into another.

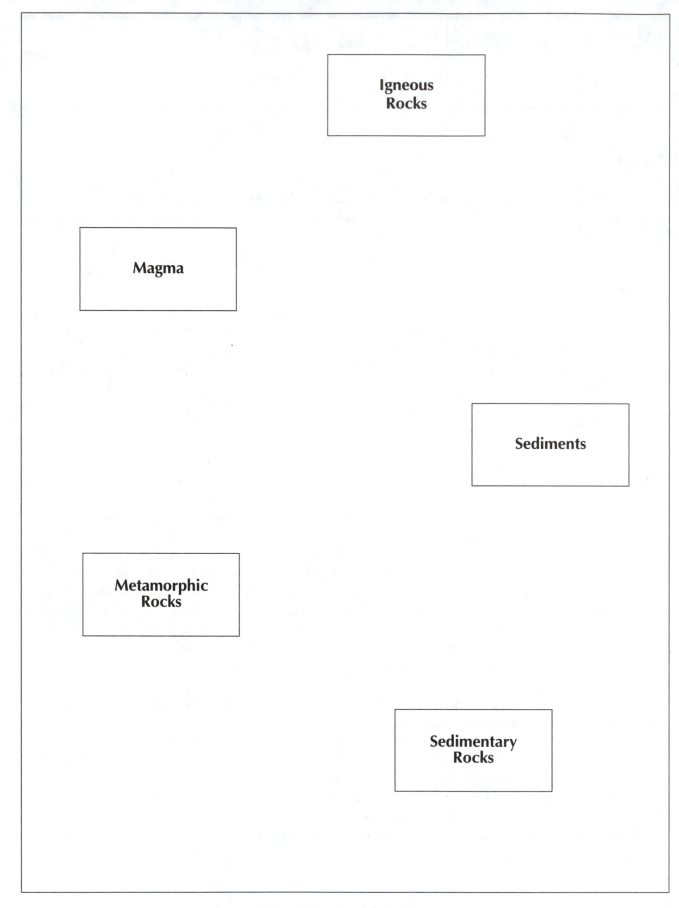

FIGURE 23–1   Rock Cycle

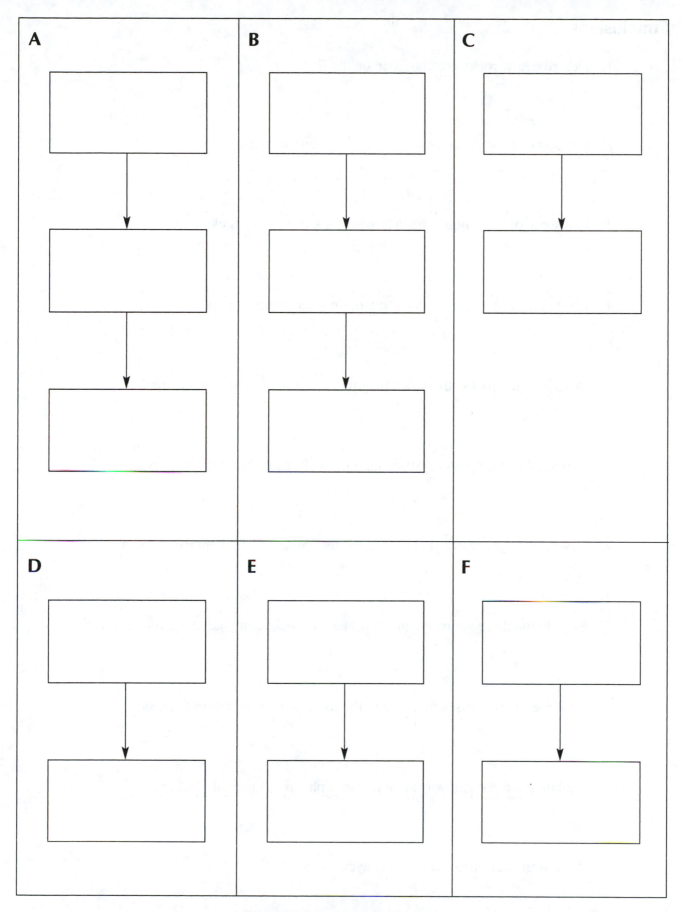

FIGURE 23–2   Rock Processes

Conclusions

1. Describe how rocks are classified on the Earth.

2. List some of the unique characteristics of igneous rocks.

3. List some of the unique characteristics of sedimentary rocks.

4. List some of the unique characteristics of metamorphic rocks.

5. Describe the process by which magma or lava forms into igneous rock.

6. Explain the three pathways that igneous rocks can take in the rock cycle.

7. Describe the processes that lead to the formation of sedimentary rocks.

8. Explain the three pathways that sedimentary rocks can take in the rock cycle.

9. Describe the processes that lead to the formation of metamorphic rocks.

10. Explain the three pathways that metamorphic rocks can take in the rock cycle.

11. What is the definition of the rock cycle?

LAB 24
Soil Erosion Rates

Purpose

The purpose of this lab is to have you determine the average annual soil erosion rates associated with four different types of land use and how they affect the environment.

Materials

four bags of dried soil samples prepared by
   your instructor for each student group
colored pencils

balances or scales
graph paper or
    spreadsheet computer application

Procedure

Complete the following steps.

1. Obtain the four different soil samples from your instructor. Each of these samples represent the average annual amount of soil eroded per square foot from an unknown land use area.
2. Using a balance or scale, weigh each soil sample to the nearest tenth of a gram. Record the weight in Table 24–1.
3. Convert your weight in grams for each of the four samples into pounds by using the equivalent of one pound equals 453.6 grams. Record your answers in Table 24–1.
4. Next, convert your data on how many pounds per square foot for each sample into pounds per acre by using the equivalent of one acre equals 43,560 square feet (pounds $ft^2 \times 43,560 \ ft^2$). Record your answers in Table 24–1.
5. Use the information on land use and average soil erosion rates shown in Table 24–2 to identify the area that each unknown sample represents. Record your answers in Table 24–1.
6. Use the data from Table 24–2 and your calculated erosion rate in pounds per acre for soil samples A, B, C, and D from Table 24–1 to construct a bar graph that graphically compares your soil samples' erosion rates to the ones shown in Table 24–2. The x-axis should be labeled "Land Use Area and Unknown Samples" and the y-axis should be labeled "Annual Erosion Rate (pounds per acre)."

TABLE 24–1	Soil Erosion Rates			
Soil Sample	Weight (grams)	Weight (lbs/ft²)	Average Annual Erosion Rate (lbs/acre)	Land Use Type
A				
B				
C				
D				

TABLE 24–2	Annual Average Soil Erosion
Woodlands	752.9 pounds/acre
Crop land without soil management	15,000.2 pounds/acre
Crop land with soil management	2,614 pounds/acre
Construction sites	39,910.8 pounds/acre

Conclusions

1. Which sample did you identify to have the highest annual soil erosion rate, and what type of land use is associated with this sample?

2. Which sample did you identify to have the lowest annual soil erosion rate, and what type of land use is associated with this sample?

3. Explain why you believe there is such a great difference in soil loss between the areas that have the highest and lowest annual soil erosion rates.

4. What is the difference in erosion rates between crop land with soil management and crop land without soil management?

5. Do you believe this lab supports the practice of soil management on farms? Why or why not?

LAB 25
Soil Mineral Composition

Purpose

The purpose of this lab is to have you collect soil samples from different areas and analyze their specific mineral composition. You will also identify the type of soil it is, based on its mineral content.

Materials

enough glass tomato sauce jars with lids
  (16 or 24 ounces) for each student
  or student group

soil samples from various areas
metric rulers
tape for labels

Procedure

Complete the following steps.

1. Add enough soil to your jar to fill it up approximately half way.
2. Use tape to label the lid of your jar with the name of the specific soil sample you are analyzing.
3. Carefully fill your jar containing the soil sample with water all the way to the top.
4. Tightly close the lid of the jar, and carefully shake your soil sample to mix it thoroughly with the water.
5. Once your sample has been thoroughly mixed, place it where it will not be disturbed for 24 hours. This will allow the soil mineral particles to settle out according to their size. Any organic material present in the soil will float to the top of the water.
6. Once your soil sample has settled, carefully turn the jar so the entire sample is visible from the side. Using your ruler, measure the total height of the soil sample (excluding the water) to the nearest tenth of a centimeter. Record this measurement in Table 25–1.
7. Next, measure the thickness of the layer of sand-sized minerals that would have settled out first and collected at the bottom of the sample to the nearest tenth of a centimeter. Record this measurement in Table 25–1.
8. Measure the thickness of the silt and clay layers, and record them in Table 25–1.
9. Gather the measurements of two different soil samples from other students and add these measurements to Table 25–1.
10. Using your data on the thickness of each layer and the total height of the sample, determine the percentage that each mineral type composes in the soil. This is accomplished by the following formula: thickness of layer ÷ total height of sample × 100 = % of mineral present.
11. Record the percentages for each type of mineral for each sample in Table 25–2.
12. Using your textbook, fill in the missing information in the Soil Textural Triangle in Figure 25–1.

13. Once your instructor has checked your Soil Textural Triangle for accuracy, use the data from Table 25–2 and your Soil Textural Triangle to determine the soil type for each soil sample. Record the soil type you determined for each sample in Table 25–2.

TABLE 25–1 Soil Measurements				
Soil Sample	Total Height of Sample	Thickness of Sand	Thickness of Silt	Thickness of Clay

TABLE 25–2 Soil Percentages				
Soil Sample	Percentage Sand	Percentage Silt	Percentage Clay	Soil Type

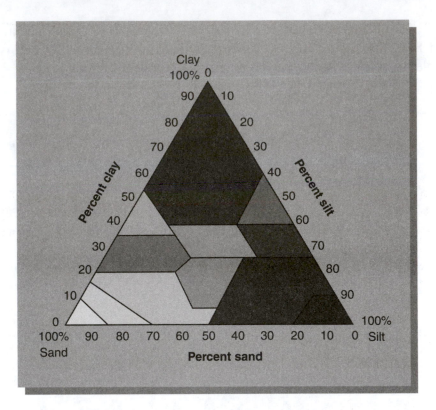

FIGURE 25–1   Soil Texture Triangle

Conclusions

1. What are the names and size ranges of mineral particles that typically make up a soil type?

2. After you shook up your soil samples in water, which mineral particle would have settled out first, and why?

3. After you shook up your soil samples in water, which mineral particle would have settled out last, and why?

4. Which soil type contains nearly equal parts of sand, silt, and clay?

5. Which soil types would have extremely poor water infiltration rates, and why?

6. Which soil type would have the greatest water infiltration rates, and why?

7. Which soil type contains approximately 60% sand, 30% silt, and 10% clay?

LAB 26
The Earth's Geologic History

Purpose

The purpose of this lab is to have you become familiar with the Earth's past history and geologic time. Using the geologic time scale, you will identify the different biologic and geologic events that have occurred during the Earth's 4.6 billion year history, and equate them to a yearly calendar.

Materials

enough copies of a yearly calendar for each student
the geologic time scale (Figure 13–4 in the textbook)

Procedure

Complete the following steps.

1. Use the information shown in the geologic time scale to help you develop a yearly calendar that illustrates the important geologic and biologic events of the Earth's history. This will be accomplished by determining time equivalents for your geologic time scale.
2. To begin, you must determine the amount of time that each day of the year represents for your time scale compared to geologic time. This is accomplished by dividing the total age of the Earth in millions of years by the amount of days in a year (4,600 million years ÷ 365 days per year). Record your answer in Table 26–1.
3. Using the number of years that represents one day for your time scale (determined in #2), calculate how many years of geologic time are equivalent to one hour of your equivalent time scale (answer from previous question ÷ 24 hours per day). Record your answer in Table 26–1.
4. Using the number of years that represents one hour in your time scale (determined in #3), calculate how many years of geologic time are equivalent to one minute of your equivalent time scale (answer from previous question ÷ 60 minutes per hour). Record your answer in Table 26–1.
5. Using the number of years that represents one hour for your time scale (determined in #4), calculate how many years of geologic time are equivalent to one second of your equivalent time scale (answer from previous question ÷ 60 seconds per minute). Record your answer in Table 26–1.
6. Once you have completed Table 26–1, use your calculated equivalent times to fill in Table 26–2 (show your work for each calculation in the space provided). Because all events that have occurred during the past 12.6 million years will have all taken place on the 31st of December in your equivalent calendar, the exact time of day that this event took place is also needed on your yearly time scale.

TABLE 26–1   Geologic Time Equivalents
One Day =
One Hour =
One Minute =
One Second =

Important Event	Actual Time (m.y.a.)	Approximate Date on Calendar	Calculations
Formation of Earth			
Oldest Rocks			
Oldest Multicellular Life			
Stromatolites			
Beginning of Cambrian Period			
First Trilobites			
Earliest Fish			
Beginning of Ordovician Period			
Taconian Orogeny			
Earliest Land Plants and Animals			
Beginning of Devonian Period			
Extinction of Armored Fish			
Earliest Amphibians			
Abundant Large Trees and Ferns			
Coal-Forming Forests			

continued

Important Event	Actual Time (m.y.a.)	Approximate Date on Calendar	Calculations
Formation of Appalachian Mountains/ Creation of Pangea			
Extinction of Trilobites			
First Dinosaurs			
Breakup of Pangea			
Earliest Birds			
First Flowering Plants			
Dinosaurs Extinct			
Continents Appear as They Do Today			
Earliest Grasses			
Mastodants, Mammoths, and Humans			
End of Last Glaciation			

TABLE 26–2   The Earth's Geologic Time Scale (continued)

Conclusions

1. What are the four basic units of time that make up the geologic time scale?

2. List all of the geologic periods that together make up the Phanerozoic eon in order from oldest to youngest.

3. Approximately what percentage of the Earth's total history did the Precambrian eon occupy?

4. Why do you think paleontologists refer to the time period that began 544 million years ago as the "Cambrian explosion"?

5. What is the name of the era during which the dinosaurs lived? How long ago did it begin and when did it end?

6. Explain how the appearance of humans on the Earth compares to its total history.

LAB 27
Radiometric Dating

Purpose

The purpose of this lab is to have you use the known radioactive decay rates of specific elements to accurately date simulated rocks, fossils, and organic remains.

Materials

250-ml beakers   sand
  (two for each student group)   graph paper

Sample Tested	Radioactive Isotope Used	Number of Half-Lives	Calculations	Age of Sample
TABLE 27–1	**Radiometric Dates of Samples**			

Procedure A

Complete the following steps.

1. Label one 250-ml beaker "A" and the other beaker "B."
2. Fill beaker A with 200 ml of sand. This will represent the amount of the radioactive isotope of carbon 14 found in the remains of a Mastondon tooth that was unearthed near the lower Hudson River in New York State.
3. Beaker B represents the amount of the daughter element, nitrogen 14, which is the product of the radioactive decay of carbon 14.
4. Record the starting amount of each element in Table 27–2.
5. Pour out half of the volume of beaker A into beaker B. This will represent one half-life. Record your data in Table 27–2.

6. Continue to pour out half of the volume of beaker A until less than 25 ml of sand remains in beaker A.
7. Using the number of half-lives it took to lower the volume of beaker A to below 20 ml, determine the age of the Mastodon tooth, knowing that the half-life of carbon 14 is 5,700 years. Record your answers and show your work in Table 27–1.

TABLE 27–2 Radioactive Decay Data for Carbon 14		
Amount of C–14 (beaker A)	Amount of N–14 (beaker B)	Number of Half-Lives
		0

Procedure B

Complete the following steps.

1. Fill beaker A with 150 ml of sand. This will represent the amount of the radioactive isotope of potassium 40 found in a sample of igneous rock from the Canadian Shield in Canada, which are the oldest rocks known to exist in North America.
2. Beaker B represents the amount of daughter element, argon 40, which is one of the products of the radioactive decay of potassium 40.
3. Record the starting amount of each element in Table 27–3.
4. Pour out half of the volume of beaker A into beaker B. This will represent one half-life. Record your data in Table 27–3.
5. Continue to pour out half of the volume of beaker A until less than 50 ml of sand remains in beaker A.
6. Using the number of half-lives it took to lower the volume of beaker A to below 50 ml, determine the age of the rock from the Canadian Shield, knowing that the half-life of potassium 40 is 1.3 billion years. Record your answers and show your work in Table 27–1.

TABLE 27–3 Radioactive Decay Data for Potassium 40		
Amount of K–40 (beaker A)	Amount of Ar–40 (beaker B)	Number of Half-Lives
		0

Procedure C

Complete the following steps.

1. Fill beaker A with 200 ml of sand. This will represent the amount of the radioactive isotope of potassium 40 found in a sample of sedimentary rock found in Australia, which are the oldest rocks known to exist on the Earth.
2. Beaker B represents the amount of daughter element, argon 40, which is one of the products of the radioactive decay of potassium 40.
3. Record the starting amount of each element in Table 27–4.
4. Pour out half of the volume of beaker A into beaker B. This will represent one half-life. Record your data in Table 27–4.
5. Continue to pour out half of the volume of beaker A until less than 50 ml of sand remains in beaker A.
6. Using the number of half-lives it took to lower the volume of beaker A to below 50 ml, determine the age of the sedimentary rock, knowing that the half-life of potassium 40 is 1.3 billion years. Record your answers and show your work in Table 27–1.

TABLE 27–4 Radioactive Decay Data for Potassium 40		
Amount of K–40 (beaker A)	Amount of Ar–40 (beaker B)	Number of Half-lives
		0

Conclusions

1. Explain why carbon 14 is good for dating the age of the remains of living things.

2. What are the four radioisotopes commonly used for radiometric dating, their half-lives, and their daughter elements?

3. How old did you determine the Mastodon tooth to be?

4. What is the approximate age of the oldest rocks found on the Earth?

5. What is the approximate age of some of the rocks that form the Canadian Shield?

6. What percentage of all the uranium that was on the Earth at the time of its formation remains on the Earth today?

LAB 28
Levels of the Earth's Atmosphere

Purpose

The purpose of this lab is to have you identify the four main layers of the Earth's atmosphere by the relationship between their unique temperatures and altitudes. You will also identify the three transitional layers within the atmosphere.

Materials

colored pencils

Procedure

The Earth's atmosphere is divided into four distinct layers, each of which is classified by its unique temperature and altitude. The areas that separate these layers are known as transitional layers.

1. Using the data from Table 28–1 and the graph paper in Figure 28–1, construct a line graph that shows the relationship between temperature and altitude in the Earth's atmosphere. The y-axis should be labeled "Altitude (km)."
2. When your graph is complete, correctly label the four levels of the atmosphere and the three transitional layers.
3. Using your colored pencils, shade in each unique layer of the atmosphere on your graph with a different color.

TABLE 28–1 Altitude and Temperature of the Earth's Atmosphere	
Altitude (kilometers)	Temperature (°C)
0	15
5	−18
10	−50
15	−57
20	−57
25	−52
30	−47
35	−37
40	−23
45	−9
50	−3
60	−26
70	−54
80	−75
90	−55
100	−5
110	50
120	125

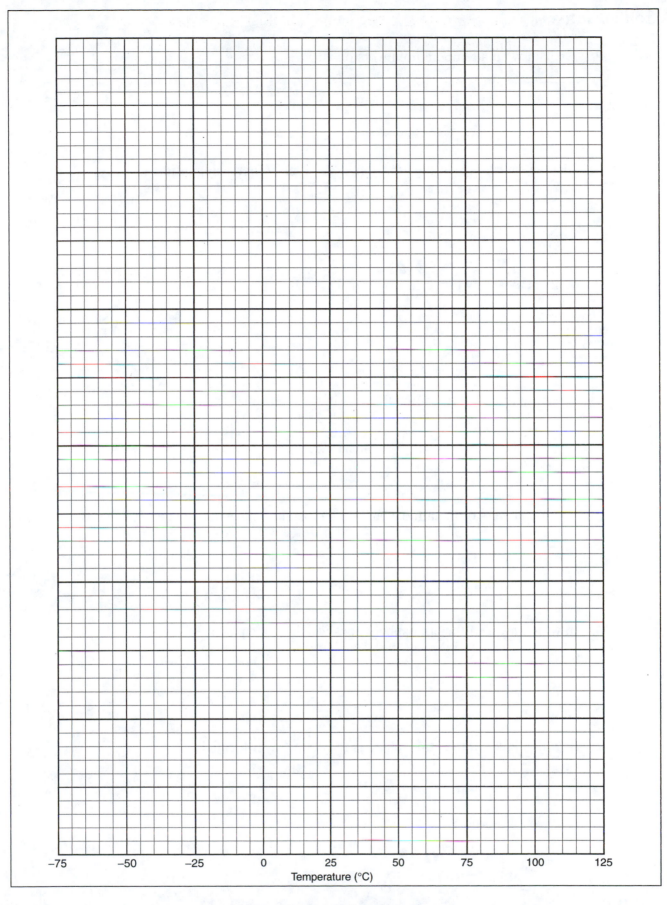

Temperature (°C)

FIGURE 28–1   Graph Paper

Conclusions

1. Explain what a temperature inversion is.

2. List the four layers of the atmosphere in order of increasing altitude.

3. What is meant by the term isothermal?

4. What are the three transitional layers in the atmosphere?

5. In which layers of the atmosphere do temperature inversions occur?

6. In which layer of the atmosphere does weather take place?

7. In which layer of the atmosphere is the ozone layer located?

LAB 29

The Interactions of Insolation
with Land and Water

Purpose

The purpose of this lab is to reveal how incoming solar radiation interacts with both land and water. Because the Earth is covered in both water and land, it is important to understand the way in which these two substances absorb and reradiate insolation, and how they affect climate.

Materials

two styrofoam cups of equal volume
two 6-inch thermometers
room temperature water
completely dry, room temperature soil
heat lamp
stopwatch
graph paper
colored pencils

SAFETY CONCERN

THE LIGHT AND THE THERMOMETERS USED IN THIS EXPERIMENT CAN GET VERY HOT! DO NOT TOUCH THEM UNTIL THEY HAVE HAD TIME TO COOL DOWN!

Procedure A

1. Fill three quarters of one of your cups with room temperature water. Then fill three quarters of the second cup with dry, room temperature soil. Try to have equal volumes of water and soil in each cup.

2. Place your cups under your light source, and adjust your light so that the bulb is approximately 18 inches above the cups (see Figure 29–1). Carefully place one thermometer in the soil and the other thermometer in the water so that the bulb of each thermometer is about 1 or 2 inches from the surface. It may be necessary to use a thermometer clip attached to a ring stand to ensure that the thermometer in suspended in the water. It is important to make sure that the bulbs of both thermometers are not touching the sides of the cups and are only recording the temperature of the soil and the water.

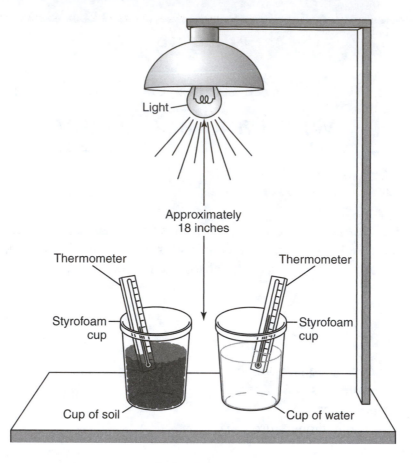

FIGURE 29–1

3. Position your cups so that they are directly under the light. DO NOT TURN ON THE LIGHT. Wait until the temperatures of both thermometers have leveled off, and record the temperature for both the soil and the water at time 0 in Table 29–1.

4. After your instructor has checked your experiment, you may turn on the light and start your stopwatch. Record the temperature for both the water and the soil once every minute for a total of 10 minutes in Table 29–1.

5. After you record the temperatures at the 10-minute mark, quickly turn off the light and carefully point it away from the cups. Continue to record the temperature of both the water and the soil for another 10 minutes. The total time for data gathering for this experiment will be 20 minutes.

6. After your experiment is complete, carefully clean up your area as directed by your instructor.

TABLE 29–1

Time	0	1	2	3	4	5	6	7	8	9	10
Soil Temperature											
Water Temperature											

Time		11	12	13	14	15	16	17	18	19	20
Soil Temperature											
Water Temperature											

Procedure B

Use the data from Table 29–1 to construct a dual line graph that shows the relationship between the heating and cooling of soil and water over a 20-minute time period. The x-axis should be labeled "Time," and the y-axis should be labeled "Temperature." Use a different colored pencil for both the soil and water, and make a key for your graph.

Procedure C

Use the data on the actual average monthly air temperature for Philadelphia and the average monthly ocean temperature off Atlantic City in Table 29–2 to create a dual line graph similar to the one you made in Procedure B. The x-axis should be labeled "Month," and the y-axis should be labeled "Temperature." Use a different colored pencil for both the ocean temperature and the air temperature over land. Make a key for your graph.

TABLE 29–2
Average Monthly Air Temperature (°C), Philadelphia, PA

Jan	Feb	Mar	Apr	May	June	July	Aug	Sep	Oct	Nov	Dec
−1	1	6	11	17	22	25	24	20	13	8	3

Average Monthly Ocean Temperature (°C), Atlantic City, NJ

Jan	Feb	Mar	Apr	May	June	July	Aug	Sep	Oct	Nov	Dec
3	2	6	9	13	18	21	22	21	16	12	7

Conclusions

1. Approximately what percentage of the Earth's surface is covered by water?

2. Approximately what percentage of the Earth's surface is covered by land?

3. What is the process called when radiation is taken in by a substance?

4. Explain what change took place during the experiment that indicated that radiation was being taken in by both the soil and the water during your experiment.

5. What is the process called when radiation is given off by a substance?

6. Explain how you could tell that radiation was given off by both the soil and the water during your experiment.

7. Which substance heated more rapidly during your experiment?

8. Which substance cooled more rapidly during your experiment?

9. Describe the difference in how water and soil absorb and reradiate energy.

10. Calculate the rate of temperature change for both soil and water during the first 10 minutes of the experiment. Show your work!

11. Calculate the rate of temperature change for both the soil and water for the final 10 minutes of your experiment. Show your work!

12. How did the graph of the actual air and water temperatures for Philadelphia and Atlantic City compare to your experimental values?

13. Explain how a coastal climate's temperature during the day would differ from that of an inland climate.

LAB 30

Terrestrial Surface Properties and the Absorption of Insolation

Purpose

The purpose of this lab is to investigate the way in which different terrestrial surfaces on Earth interact with incoming solar radiation.

Materials

~ 150 milliliters of dry instant coffee crystals
~ 150 milliliters of dry pink lemonade crystals
~ 150 milliliters of dry white sugar crystals
three styrofoam cups
four 6-inch thermometers
stopwatch
heat lamp
graph paper
colored pencils

SAFETY CONCERN
THE LIGHT AND THE THERMOMETERS USED IN THIS EXPERIMENT CAN GET VERY HOT! DO NOT TOUCH THEM UNTIL THEY HAVE HAD TIME TO COOL DOWN!

Procedure A

1. Carefully fill one styrofoam cup about two-thirds to three-quarters full with the coffee crystals. Repeat for the lemonade crystals and sugar. Next, carefully slide the bulb of a thermometer into each substance, approximately 1 to 2 inches deep. Your experiment should resemble Figure 30–1.

2. Carefully position the three cups underneath your light (see Figure 30–1). DO NOT TURN ON YOUR LIGHT. Make sure that your light is approximately 18 inches above the cups, and that each cup is receiving equal amounts of light. Wait a few minutes for all three of your thermometers to become stable, and record each of the three temperatures at time 0 in the correct location in Table 30–1.

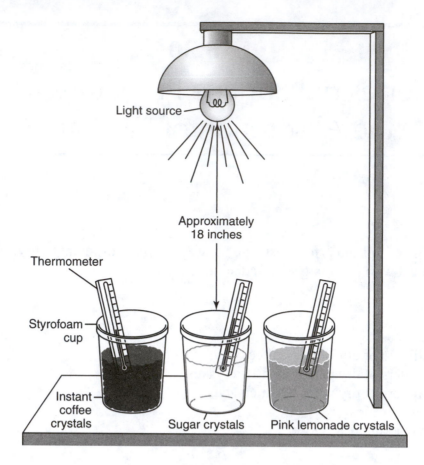

FIGURE 30–1

TABLE 30–1											
Time	0	1	2	3	4	5	6	7	8	9	10
Instant Coffee Temperature											
Sugar Temperature											
Pink Lemonade Temperature											
Time	11	12	13	14	15	16	17	18	19	20	
Instant Coffee Temperature											
Sugar Temperature											
Pink Lemonade Temperature											

3. After your instructor has checked your experiment, turn on your light and start your stopwatch. Record the temperature of each material once every minute for a total of 10 minutes in Table 30–1.

4. After 10 minutes, turn off your light and carefully point it away from the cups. Continue to record the temperature for each material every minute for another 10 minutes in Table 30–1.

5. Wait for your experiment to completely cool down before cleaning up. Follow the directions for cleanup as outlined by your instructor.

Procedure B

1. Use the data from Table 30–1 to construct a multiple line graph that shows the relationship between the heating and cooling of all three materials over the 20-minute time period. The x-axis should be labeled "Time," and the y-axis should be labeled "Temperature." Use a different colored pencil for each material, and make a key for your graph.

Conclusions

1. Which of the three materials heated up the fastest? Cooled off the fastest?

2. Which of the three materials heated up the slowest? Cooled off the slowest?

3. Calculate the rate of temperature change for each of the three materials during the first 10 minutes of your experiment. Show your work.

4. Calculate the rate of temperature change for each of the three materials during the last 10 minutes of your experiment. Show your work.

5. What change occurred during the experiment that helped you determine whether a material was a good absorber of radiation?

6. What happened during your experiment that helped you determine whether a material was a good reflector?

7. What are two possible sources of error that might have affected the outcome of this experiment?

8. Describe the properties associated with a material that is a good absorber of radiation.

9. Describe the properties associated with a material that is a good reflector of radiation.

10. What material on the Earth would be an excellent reflector of insolation?

11. Which terrestrial surface would be a better absorber of insolation: a dark green forest or a light colored sandy desert? Why?

12. Describe the general relationship between the absorption and reradiation of energy by earth materials.

LAB 31
Angle of Insolation

Purpose

The purpose of this lab is to have you determine the changes in the angle of insolation that occur throughout the year at different latitudes on the Earth. You will then use this information to identify the relationship between temperature and angle of insolation, and how this affects the seasons.

Materials

protractor

graph paper or
computer spreadsheet application

Procedure

Complete the following steps.

1. Using a protractor, determine the angle at which incoming solar radiation (insolation) is striking the Earth's surface at noon at the three latitude locations shown in diagram A, Figure 31–1. Record the angle to the nearest whole degree for each location in Table 31–1.
2. Use the above procedure to determine the angles of insolation for all of the latitude locations for diagrams B and C, Figure 31–1. Record the angles for each location in Table 31–1.
3. Using the data on the average monthly temperature for 42 degrees north latitude provided in Table 31–2, create a line graph that shows the relationship between average temperature and month during the year. The x-axis should be labeled "Month" and the y-axis should be labeled "Temperature (°F)."

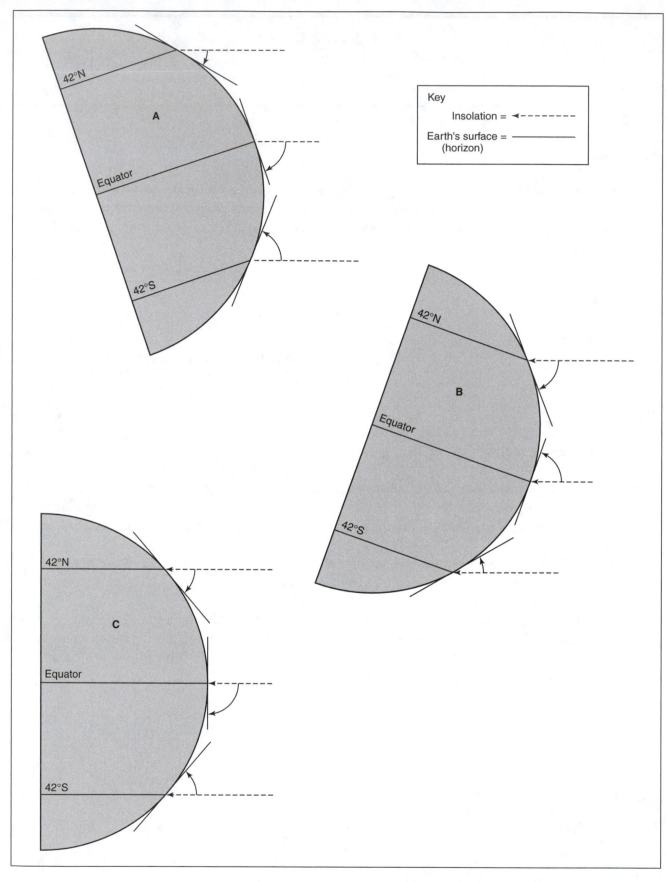

Key

Insolation = ◄ - - - - - - - -

Earth's surface = ————
(horizon)

FIGURE 31–1   Insolation Angles

TABLE 31–1   Insolation Diagram

Diagram A

Latitude	Angle of Insolation
42° North	
Equator	
42° South	

Diagram B

Latitude	Angle of Insolation
42° North	
Equator	
42° South	

Diagram C

Latitude	Angle of Insolation
42° North	
Equator	
42° South	

TABLE 31–2	Average Temperatuare Table
Month	Average Temperature at 42° North (°F)
January	20.7
February	27.6
March	40.1
April	45.3
May	59.3
June	65.8
July	67.6
August	68.5
September	59.3
October	50.0
November	38.0
December	22.1

Conclusions

1. Using your data on the angle of insolation for diagram A, what season of the year do you believe this diagram represents?

2. Using your data on the angle of insolation for diagram B, what season of the year do you believe this diagram represents?

3. Using your data on the angle of insolation for diagram C, what seasons of the year do you believe this diagram represents?

4. What is the lowest angle of insolation that you determined the equator receives throughout the year, and during what season does it occur?

5. What is the lowest angle of insolation received at 42 degrees north latitude that you determined throughout the year, and during what season does it occur?

6. What is the highest angle of insolation received at 42 degrees north latitude that you determined throughout the year, and during what season does it occur?

7. Using your data on angle of isolation during spring and fall at different latitudes, what is the general relationship between angle of insolation and latitude location on the Earth?

8. Using your data, describe the relationship between the season of the year in the Northern Hemisphere and the angle of insolation.

9. Using your line graph showing average monthly temperature, describe the relationship between the angle of insolation and average temperature on the Earth.

LAB 32
Insolation, Latitude, and Temperature

Purpose

The purpose of this lab is to show you how the curvature of the Earth affects the angle at which insolation is striking the planet's surface, therefore affecting its surface temperature. This experiment also reveals the relationship between latitude location on the Earth and surface temperature.

Materials

classroom globe (at least 30 cm in diameter)
three small thermometers
light source
tape
stopwatch
colored pencils
graph paper

SAFETY CONCERN
THE LIGHT AND THE THERMOMETERS USED IN THIS EXPERIMENT CAN GET VERY HOT! DO NOT TOUCH THEM UNTIL THEY HAVE HAD TIME TO COOL DOWN!

Procedure A

1. Using Figure 32–1 as a guide, carefully tape the bulb of each thermometer so that it is positioned over the prime meridian at the following latitude locations: the equator (0 degrees), 45 degrees north, and the North Pole (90 degrees north). Place the tape between the bulb and the current temperature shown on each thermometer; this will enable you to read the temperature during your experiment. Do not cover the bulbs of the thermometers with tape.

2. Carefully place your globe so that it is approximately 2 feet from your light source. Have your instructor help you set up the globe so that the prime meridian is in an equinox position (the North Pole is not tilted toward or away from the light source). Adjust your light so that the bulb is pointed at the equator on your globe (see Figure 32–1). DO NOT TURN ON YOUR LIGHT UNTIL YOUR INSTRUCTOR HAS CHECKED YOUR EXPERIMENT. After your instructor has checked your setup, record the temperature for time 0 for each thermometer in the correct place in Table 32–1. After you have recorded the start temperatures for each location, you may turn on your light

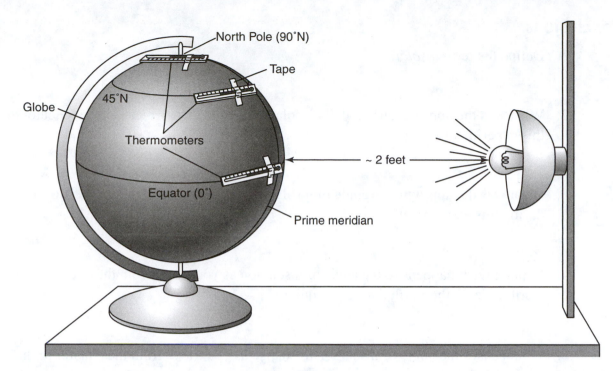

FIGURE 32–1

TABLE 32–1																
Time	0	1	2	3	4	5	6	7	8	9	10	11	12	13	14	15
Temperature at the Equator (0°)																
Temperature at 45° North																
Temperature at the North Pole (90°N)																

and start your stopwatch. You will then record the temperature for each thermometer once every minute for a total of 15 minutes. During your experiment make sure not to touch either the thermometers or the light because they can get very hot!

3. After your experiment is complete, carefully turn off your light. Do not touch either the light or the thermometers until they have had time to cool down.

Procedure B

1. Use the data from Table 32–1 to construct a multiple line graph that shows the relationship between latitude location and temperature on your globe. The x-axis should be labeled "Time," and the y-axis should be labeled "Temperature." Use a different colored pencil for each latitude location, and make a key for your graph.

Conclusions

1. Define the term *insolation*.

2. What was the approximate angle of insolation for the thermometer on the equator during this experiment?

3. What was the approximate angle of insolation for the thermometer on the North Pole during this experiment?

4. Explain what happens to the angle of insolation as you travel from the equator to the North Pole on the Earth during an equinox.

5. What causes the angle of insolation to change as you travel from the equator to the poles on the Earth?

6. Why was it important to tape the bulbs of the thermometers along the prime meridian?

7. Which latitude location in your experiment experienced the highest temperature. Why?

8. Which latitude location in your experiment experienced the lowest temperature? Why?

9. Calculate the rate of temperature change for each latitude location for the first 10 minutes of the experiment. Show your work.

10. Describe the relationship between latitude location and surface temperature on the Earth.

11. Explain how this experiment helped to prove why it is usually cooler in the higher latitudes and warmer near the equator.

LAB 33
The Greenhouse Effect and Surface Temperature

Purpose

The purpose of this lab is to illustrate the concept known as the greenhouse effect and how it helps regulate surface temperature on the Earth.

Materials

one 300-milliliter glass beaker
two 6-inch thermometers
~ 50 milliliters of dry instant coffee crystals
8.5″ × 11″ white piece of paper
light source
stopwatch
colored pencils
graph paper

SAFETY CONCERN
THE LIGHT AND THE THERMOMETERS USED IN THIS EXPERIMENT CAN GET VERY HOT! DO NOT TOUCH THEM UNTIL THEY HAVE HAD TIME TO COOL DOWN!

Procedure A

1. Make two small piles of instant coffee (approximately 25 milliliters each) on the white piece of paper. Carefully lay the bulb of each thermometer on top of each pile (see Figure 33–1). Cover one thermometer with the 300-milliliters beaker.

2. Position your light source over the two piles of coffee, making sure that it is approximately 18 inches above the thermometers (see Figure 33–1). DO NOT TURN YOUR LIGHT ON. Wait until the thermometers have leveled off, and then record the temperature for each pile at time 0 in Table 33–1.

3. After your instructor has checked your experiment, turn on your light and start your stopwatch. Record the temperature for both piles once every minute for 10 minutes in Table 33–1. DO NOT TOUCH THE BEAKER DURING THIS EXPERIMENT BECAUSE IT MAY GET VERY HOT!

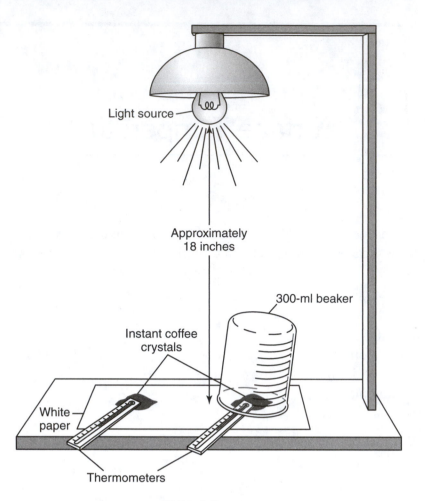

Light source

Approximately
18 inches

300-ml beaker

Instant coffee
crystals

White
paper

Thermometers

FIGURE 33–1

TABLE 33–1											
Time	0	1	2	3	4	5	6	7	8	9	10
Open Coffee											
Covered Coffee											
Time		11	12	13	14	15	16	17	18	19	20
Open Coffee											
Covered Coffee											

4. After the first 10 minutes, turn off your light and carefully point it away from the piles. Continue to record the temperature of each pile of coffee once every minute for another 10 minutes in Table 33–1.

5. Make sure you allow the experiment to cool down completely before you clean up. Follow your instructor's directions for proper cleanup procedure.

Procedure B

1. Use the data from Table 33–1 to construct a dual line graph that shows the relationship between the heating and cooling of the open coffee and the one covered with the beaker. The x-axis should be labeled "Time," and the y-axis should be labeled "Temperature." Use a different colored pencil for each beaker, and make a key for your graph.

Conclusions

1. Calculate the rate of heating for both piles of coffee during the first 10 minutes of the experiment. Show your work!

2. Calculate the rate of cooling for both piles of coffee during the last 10 minutes of the experiment. Show your work!

3. Which thermometer heated up more quickly? Why?

4. Which thermometer cooled off more quickly? Why?

5. Which retained the most heat, the open thermometer or the covered thermometer? Why?

6. What role did the 300-milliliter beaker play in this experiment?

7. Explain the process known as the greenhouse effect.

8. Which gases in the Earth's atmosphere act like the 300-milliliter beaker in the experiment?

9. What would happen to the Earth's surface temperature if it had no atmosphere?

10. What would happen to the Earth's surface temperature if it contained a higher concentration of greenhouse gases?

11. Describe how the planets Mercury, Venus, Earth, and Mars are affected by the greenhouse effect.

LAB 34
Atmospheric Pressure and Temperature

Purpose

The purpose of this lab is to have you identify the relationship between atmospheric temperature and pressure near the surface of the Earth. You will also use the technique of drawing isotherms for the analysis of surface temperature on surface weather maps.

Materials

graph paper or
   computer spreadsheet application

colored pencils

Procedure A

Using the data in Table 34–1, create a line graph showing the temperature and barometric pressure for one 24-hour period. Plot your data for temperature in one colored pencil and the data for barometric pressure in another colored pencil. Label both lines on your graph. The x-axis should be labeled "Time of Day," the primary y-axis should be labeled "Barometric Pressure (millibars)," and the secondary y-axis should be labeled "Temperature (°F)."

Then create a second line graph using the data in Table 34–2 for a different day's temperature and barometric pressure. Set up this graph in the same way as your first.

TABLE 34–1 Temperature and Barometric Pressure		
Time	Temp. Out	Bar
12:00a	40.1	1025.6
1:00a	40.8	1023.4
2:00a	40.2	1023.2
3:00a	40.1	1021.2
4:00a	41.3	1020.3
5:00a	42.3	1019.2
6:00a	42.3	1018.0
7:00a	40.6	1016.9
8:00a	41.2	1016.3
9:00a	42.9	1015.7
10:00a	44.9	1014.3
11:00a	46.4	1013.4
12:00p	48.6	1011.7
1:00p	48.9	1011.1
2:00p	50.2	1011.0
3:00p	48.3	1010.9
4:00p	48.2	1011.2
5:00p	48.8	1010.9
6:00p	47.7	1010.9
7:00p	48.2	1010.9
8:00p	48.2	1010.8
9:00p	48.5	1011.8
10:00p	48.9	1012.2
11:00p	50.1	1012.4
12:00p	50.1	1012.9

TABLE 34–2 Temperature and Barometric Pressure		
Time	Temp. Out	Bar
12:00a	38.9	1001.5
1:00a	39.6	1000.7
2:00a	39.2	999.5
3:00a	39.3	998.2
4:00a	39.2	997.6
5:00a	38.2	997.5
6:00a	36.9	995.7
7:00a	36.3	995.3
8:00a	36.5	995.4
9:00a	36.2	996.0
10:00a	36.8	995.4
11:00a	38.3	993.9
12:00p	37.5	993.5
1:00p	35.9	994.4
2:00p	35.5	995.6
3:00p	34.7	996.4
4:00p	32.7	998.1
5:00p	32.4	999.9
6:00p	31.9	1001.8
7:00p	31.2	1003.0
8:00p	30.3	1004.6
9:00p	29.4	1006.3
10:00p	29.4	1007.9
11:00p	29.1	1009.1
12:00p	28.2	1010.0

Procedure B

Using the techniques of isotherm analysis introduced to you by your instructor, draw isotherms at 10-degree intervals on the two surface temperature maps in Figures 34–1 and 34–2.

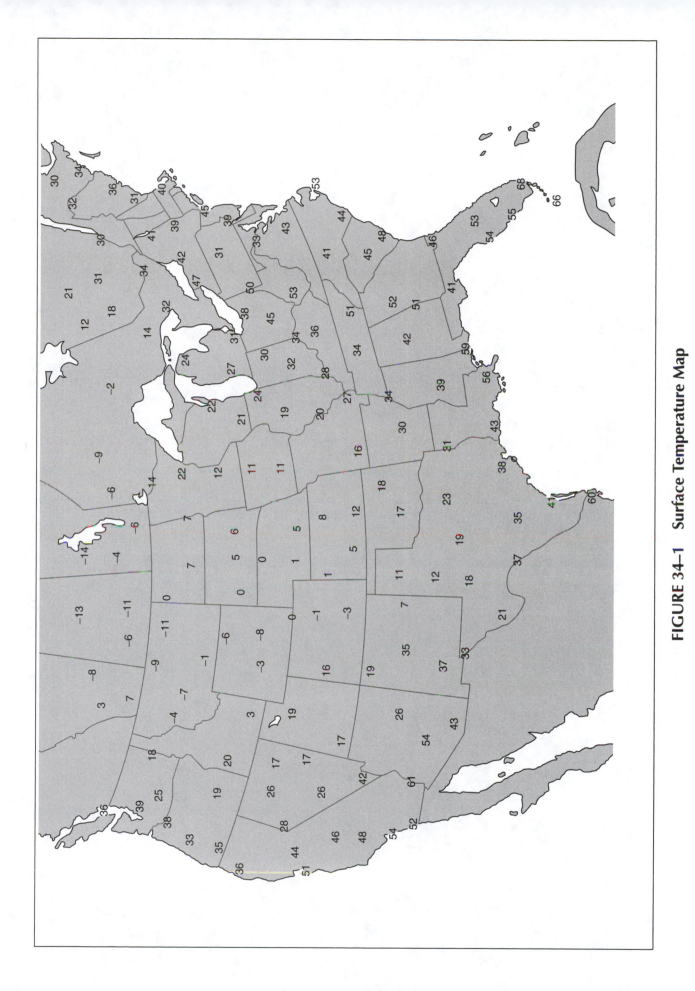

FIGURE 34–1   Surface Temperature Map

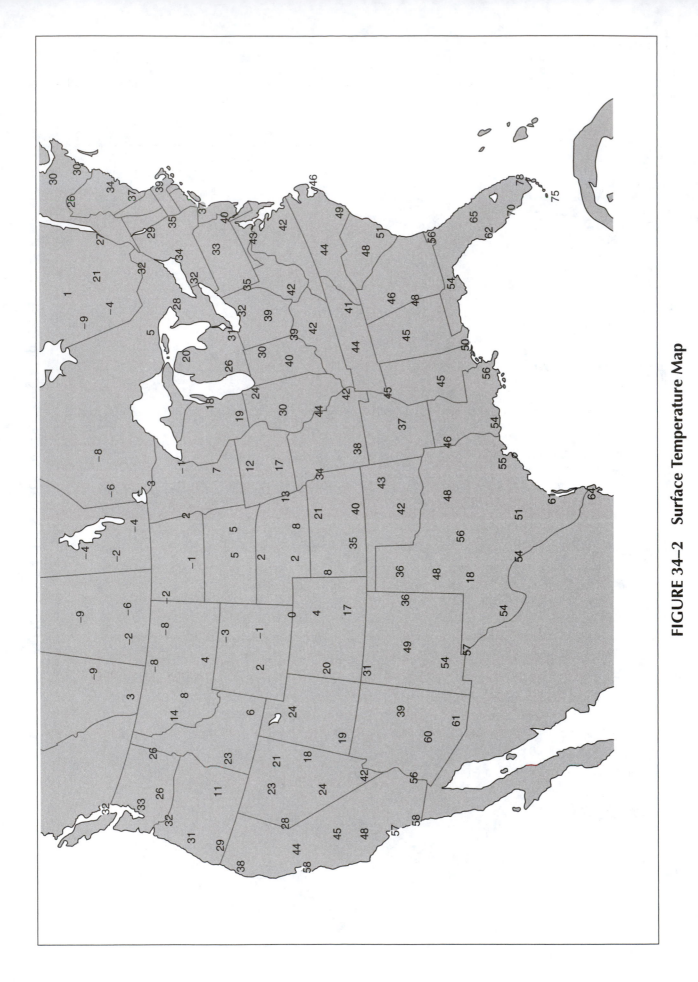

FIGURE 34-2 Surface Temperature Map

Procedure C

Using information from your textbook and help from your instructor, fill in the characteristics associated with both high- and low-pressure systems at the Earth's surface in Table 34–3.

TABLE 34–3 Atmospheric Pressure Systems		
Pressure		
Temperature		
Wind Rotation (Northern Hemisphere)		
Moisture		
Weather Map Symbol		
Rising or Sinking Air		

Conclusions

1. Using your graphs from Procedure A, describe the general relationship between temperature and barometric pressure near the Earth's surface.

2. A decrease in atmospheric pressure would signal what possible change in the current temperature?

3. A decrease in the present air temperature would signal what type of change in atmospheric pressure?

4. Using the data from Table 34–1, what was the rate of temperature change between 12:00 A.M. and 2:00 P.M. (rate of change = change in value ÷ change in time)? Show your work.

5. Using the data from Table 34–1, what was the rate of change of the barometric pressure between 12:00 A.M. and 2:00 P.M. (rate of change = change in value ÷ change in time)? Show your work.

6. Using the data from Table 34–2, what was the rate of temperature change between 11:00 A.M. and 12:00 P.M. (rate of change = change in value ÷ change in time)? Show your work.

7. Using the data from Table 34–2, what was the rate of change of the barometric pressure between 11:00 A.M. and 12:00 P.M. (rate of change = change in value ÷ change in time)? Show your work.

8. What are lines that connect equal values of temperature on a weather map called?

9. When drawing isolines on a map, what are three rules you should always follow?

10. Describe the direction of wind rotation associated with a low-pressure system forming in South America.

LAB 35

Dew Point Temperature, Relative Humidity, and Cloud Formation

Purpose

The purpose of this activity is for you to calculate the dew-point temperature and the relative humidity by using wet-bulb and dry-bulb temperatures. You will also determine the height at which clouds will form, based on the current dew-point temperature.

Procedure A

The dew point temperature can be calculated by using the difference between the wet-bulb and dry-bulb temperature, and dew-point temperatures. The relative humidity of the air can be calculated by using the difference between the wet-bulb and dry-bulb temperature, and relative humidity.

1. Using the data from Table 35–1 showing the wet-bulb and dry-bulb temperatures for ten weather observations, and the dew-point temperatures in Table 35–2, calculate the dew point for each observation and record them in the spaces provided in Table 35–1. If you have a psychrometer in your classroom, and with help from your instructor, you can also determine the present wet-bulb and dry-bulb temperatures for the outside air. Record them in Space 11 in Table 35–1.
2. Using the data from Table 35–1 on the wet-bulb and dry-bulb temperatures for ten weather observations, and the relative humidity readings in Table 35–3, calculate the relative humidity for each observation and record them in the spaces provided in Table 35–1.

TABLE 35–1 Wet- and Dry-Bulb Temperature Chart					
Observation	Dry-Bulb Temp. (°C)	Web-Bulb Temp. (°C)	Wet-Bulb Depression	Dew-Point Temp. (°C)	Relative Humidity
1	28	18			
2	24	16			
3	6	3			
4	20	19			
5	18	13			
6	26	15			
7	−2	0			
8	10	6			
9	19	14			
10	28	14			
11					

TABLE 35–2   Dew-Point Temperatures (°C)

Dry-Bulb Temperature (°C)	Difference Between Wet-Bulb and Dry-Bulb Temperatures (°C)															
	0	1	2	3	4	5	6	7	8	9	10	11	12	13	14	15
–20	–20	–33														
–18	–18	–28														
–16	–16	–24														
–14	–14	–21	–36													
–12	–12	–18	–28													
–10	–10	–14	–22													
–8	–8	–12	–18	–29												
–6	–6	–10	–14	–22												
–4	–4	–7	–12	–17	–29											
–2	–2	–5	–8	–13	–20											
0	0	–3	–6	–9	–15	–24										
2	2	–1	–3	–6	–11	–17										
4	4	1	–1	–4	–7	–11	–19									
6	6	4	1	–1	–4	–7	–13	–21								
8	8	6	3	1	–2	–5	–9	–14								
10	10	8	6	4	1	–2	–5	–9	–14	–28						
12	12	10	8	6	4	1	–2	–5	–9	–16						
14	14	12	11	9	6	4	1	–2	–5	–10	–17					
16	16	14	13	11	9	7	4	1	–1	–6	–10	–17				
18	18	16	15	13	11	9	7	4	2	–2	–5	–10	–19			
20	20	19	17	15	14	12	10	7	4	2	–2	–5	–10	–19		
22	22	21	19	17	16	14	12	10	8	5	3	–1	–5	–10	–19	
24	24	23	21	20	18	16	14	12	10	8	6	2	–1	–5	–10	–18
26	26	25	23	22	20	18	17	15	13	11	9	6	3	0	–4	–9
28	28	27	25	24	22	21	19	17	16	14	11	9	7	4	1	–3
30	30	29	27	26	24	23	21	19	18	16	14	12	10	8	5	1

TABLE 35–3   Relative Humidity (%)

Dry-Bulb Temperature (°C)	Difference Between Wet-Bulb and Dry-Bulb Temperatures (°C)															
	0	1	2	3	4	5	6	7	8	9	10	11	12	13	14	15
–20	100	28														
–18	100	40														
–16	100	48														
–14	100	55	11													
–12	100	61	23													
–10	100	66	33													
–8	100	71	41	13												
–6	100	73	48	20												
–4	100	77	54	32	11											
–2	100	79	58	37	20	1										
0	100	81	63	45	28	11										
2	100	83	67	51	36	20	6									
4	100	85	70	56	42	27	14									
6	100	86	72	59	46	35	22	10								
8	100	87	74	62	51	39	28	17	6							
10	100	88	76	65	54	43	33	24	13	4						
12	100	88	78	67	57	48	38	28	19	10	2					
14	100	89	79	69	60	50	41	33	25	16	8	1				
16	100	90	80	71	62	54	45	37	29	21	14	7	1			
18	100	91	81	72	64	56	48	40	33	26	19	12	6			
20	100	91	82	74	66	58	51	44	36	30	23	17	11	5		
22	100	92	83	75	68	60	53	46	40	33	27	21	15	10	4	
24	100	92	84	76	69	62	55	49	42	36	30	25	20	14	9	4
26	100	92	85	77	70	64	57	51	45	39	34	28	23	18	13	9
28	100	93	86	78	71	65	59	53	47	42	36	31	26	21	17	12
30	100	93	86	79	72	66	61	55	49	44	39	34	29	25	20	16

Procedure B

The altitude at which clouds will form can be calculated by using the current dew-point temperature for the surrounding air. Using the data on the dew point for the ten weather observations you determined in Table 35–1 and the Cloud Base Altitude chart in Figure 35–1, calculate the height at which clouds will form and record them in the spaces provided in Table 35–4. The cloud base altitude can be determined by using the air temperature and dew-point temperature. Trace the solid air temperature line and the dashed dew-point temperature line with both values to be determined, and where the two converge, read over to the height of the cloud base.

TABLE 35–4 Cloud Base Altitude Table											
	1	2	3	4	5	6	7	8	9	10	11
Air Temp. (°C)											
Dew-Point Temp. (°C)											
Cloud Base Altitude (km)											

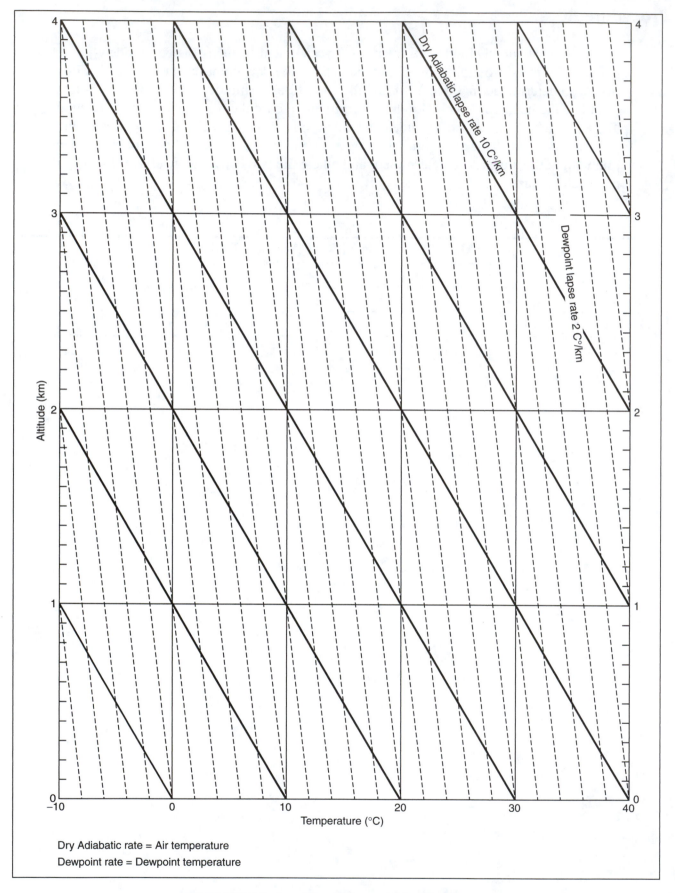

Dry Adiabatic lapse rate 10 C°/km

Dewpoint lapse rate 2 C°/km

Dry Adiabatic rate = Air temperature

Dewpoint rate = Dewpoint temperature

FIGURE 35–1   Cloud Base Altitude Chart

Conclusions

1. What information do you need to calculate the dew-point temperature or relative humidity?

2. How do you calculate the wet-bulb depression?

3. Explain what happens to the humidity of the air as the temperature becomes closer to the dew point.

4. What happens to the height of cloud formation as the difference between the dew-point and air temperature increases?

LAB 36
Pressure Gradient, Wind, and Air Masses

Purpose

The purpose of this lab is to have you identify the relationship between winds on the Earth and atmospheric pressure. You will also be able to identify the global patterns of planetary scale winds.

Materials

colored pencils

Procedure A

Complete the following steps.

1. On the blank diagram of the Earth in Figure 36–1, label the location of the equator, the North Pole, South Pole, and the following lines of latitude: 30 degrees north, 60 degrees north, 30 degrees south, and 60 degrees south.
2. Using your textbook, add the locations of the equatorial and subpolar low pressure centers with a red "L."
3. Next, add the locations of the subtropical and polar high pressure centers with a blue "H."
4. Finally, add purple arrows that show the direction of, and label, the following planetary scale winds: the prevailing westerlies, the northeast trade winds, the southeast trade winds, and the polar easterlies.

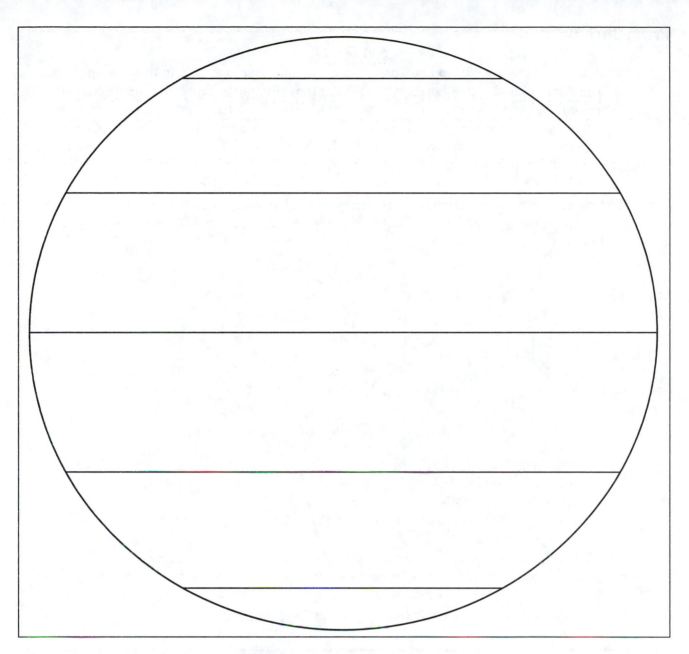

FIGURE 36–1   Global Wind Patterns

Procedure B

Pressure gradient force, also known as wind, is an important variable in the atmosphere that helps to define weather on Earth. By analyzing the differences in pressure between two points on the Earth's surface, it is often possible to predict local wind direction and speed. Using the surface weather map (Figure 36–2), calculate the pressure gradient in millibars per mile between the points shown in Table 36–1. Show your calculations and record your answers in the spaces provided in Table 36–1.

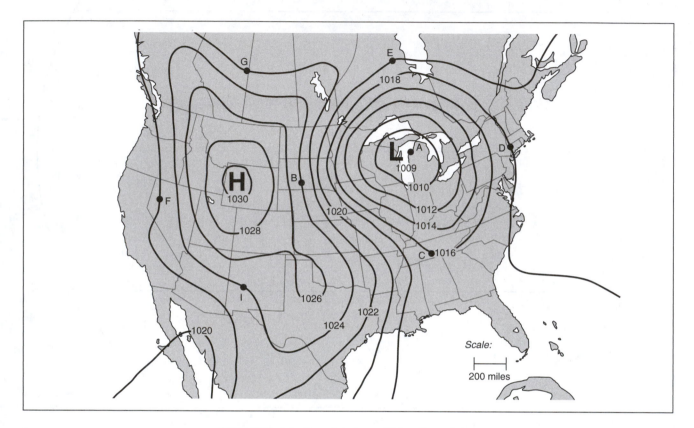

FIGURE 36–2   Surface Weather Map

TABLE 36–1 Pressure Gradient				
Points on Map	Change in Pressure (mb)	Distance (miles)	Calculations	Pressure Gradient
A–B				
A–C				
A–D				
A–E				
H–F				
H–G				
H–I				

Procedure C

Complete the following steps.

1. Using your textbook, fill in the information in Table 36–2 in the characteristics of the five types of air masses.
2. On the blank map of North America in Figure 36–3, label the source areas for the five air mass types you identified in Table 36–2.

TABLE 36–2 Air Masses			
Symbol	Name	Where It Forms	Characteristics (Temperature and Moisture)

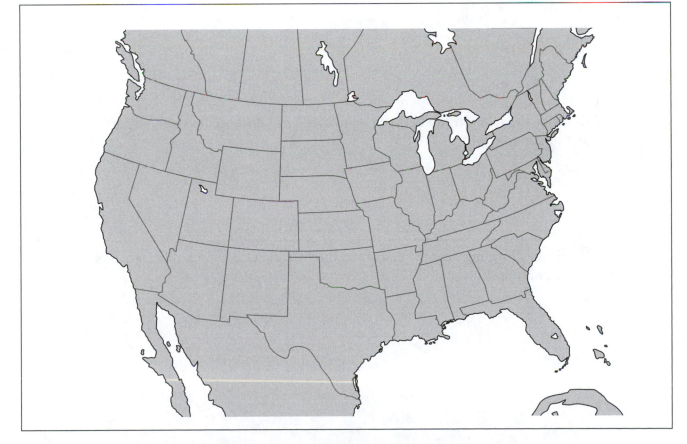

FIGURE 36–3   Air Mass Source Regions

Procedure D

Using the surface weather map in Figure 36–4, analyze the surface wind patterns by drawing arrows across each weather station through the wind vane, showing the direction the wind is blowing.

Conclusions

1. Explain the process by which planetary scale winds form on the Earth.

2. What causes the trade winds to blow from the northeast and not directly from the north between 30 degrees north latitude and the equator?

3. Using the information you calculated in Procedure B, which area of the country will most likely experience the strongest winds, and from what direction will they be coming?

4. Describe the relationship between wind speed and distance between isobars on a pressure map.

5. Hurricanes forming near the equator in the Atlantic Ocean probably form from which type of air mass?

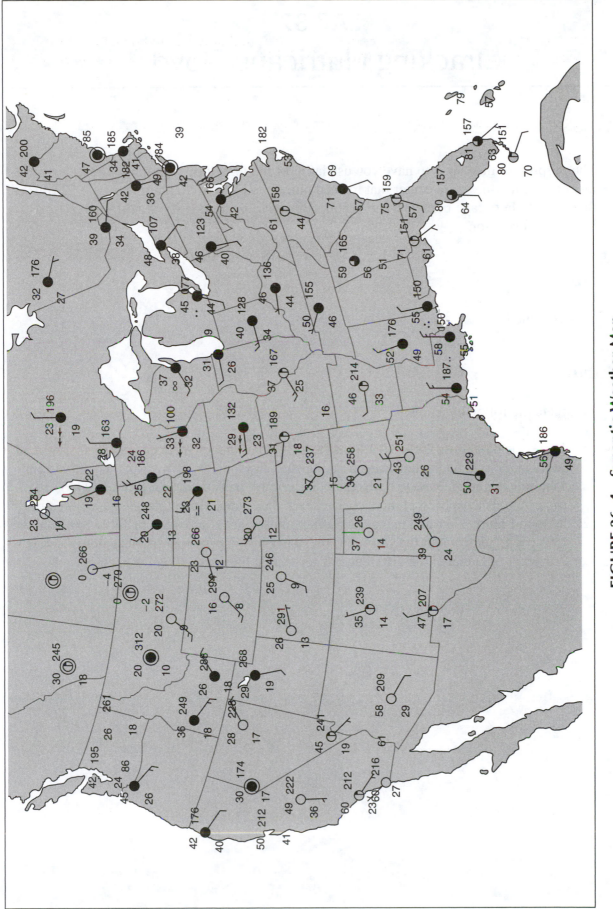

FIGURE 36—4 Synoptic Weather Map

LAB 37
Tracking Hurricane Floyd

Purpose

The purpose of this lab is to have you use latitude and longitude location to track the movement of the low-pressure center that formed hurricane Floyd in September of 1999. You will then use the hurricane track and the Saffir/Simpson scale to identify the different stages of hurricane formation.

Materials

colored pencils

Procedure

Complete the following steps.

1. Using the data from Table 37–1, plot the location of the low-pressure center that formed hurricane Floyd on the hurricane tracking chart in Figure 37–1 by using its latitude and longitude location. Next to each data point, record the date for that position. Once all of the positions have been plotted, connect each data point with a line using a colored pencil.
2. Also in Figure 37–1 and using the Saffir/Simpson scale provided in Table 37–2, label the specific date when the storm became a tropical depression, tropical storm, and hurricane. Also label the different categories that hurricane Floyd was classified.

Date/Time (UTC)	Position		Pressure (mb)	Wind Speed (kt)	Stage
	Lat. (°N)	Lon. (°W)			
7/1800	14.6	45.6	1008	25	tropical depression
8/0000	15	46.9	1007	30	tropical depression
1200	15.8	49.6	1003	40	tropical storm
9/0000	16.7	52.6	1000	45	tropical storm
1200	17.3	55.1	1003	50	tropical storm
10/0000	18.3	57.2	995	60	tropical storm
1200	19.3	58.8	989	70	tropical storm
11/0000	20.8	60.4	971	80	hurricane
1200	21.9	62	962	95	hurricane
12/0000	22.7	64.1	967	85	hurricane
1200	23	66.2	955	105	hurricane
13/0000	23.4	68.7	931	125	hurricane
1200	23.9	71.4	921	135	hurricane
14/0000	24.5	74	924	115	hurricane
1200	25.4	76.3	930	105	hurricane
15/0000	27.1	77.7	933	115	hurricane
1200	29.3	78.9	943	100	hurricane
16/0000	32.1	78.7	950	90	hurricane
1200	35.7	76.8	967	70	tropical storm
17/0000	40.6	73.5	980	50	tropical storm
1200	43.3	70.6	984	45	tropical storm
18/0000	44.8	67.3	987	40	tropical storm
1200	46.6	63	992	35	tropical storm
19/0000	48	56.3	992	35	tropical storm

TABLE 37–1 Hurricane Floyd, September 1999

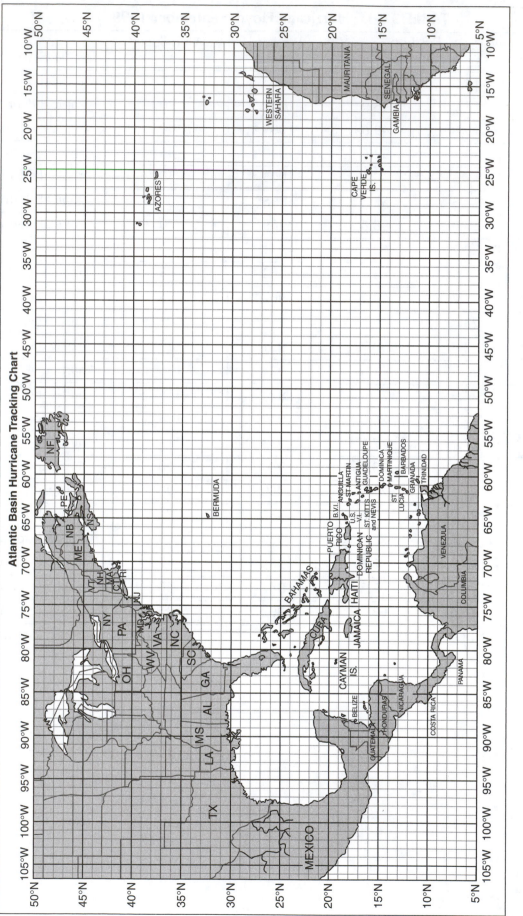

FIGURE 37–1 Atlantic Basin Hurricane Tracking Chart

TABLE 37–2 Saffir/Simpson Hurricane Scale				
Scale Number (category)	Pressure (millibars)	Winds (mph)	Storm Surge (ft)	Damage
Trop. Depression	—	<38	—	—
Tropical Storm	—	39–73	—	—
I	>979	74–95	4–5	Minimal
II	965–979	96–110	6–8	Moderate
III	945–964	111–130	9–12	Extensive
IV	920–944	131–155	13–18	Extreme
V	<920	>155	>18	Catastrophic

Conclusions:

1. At what wind speed do tropical storms become classified as hurricanes?

2. What happened to the wind speed when hurricane Floyd made landfall on the coast of North Carolina on September 16?

3. At what date and time did the wind speed reach its greatest speed?

4. At what date and time did the storm experience the lowest barometric pressure?

5. Describe the relationship between atmospheric pressure and wind speed in a hurricane.

6. During the first eight days of the storm, in what general direction was Floyd moving?

7. After September 15, what happened to the direction in which the storm was moving?

8. Explain what might have caused the direction the storm was traveling as determined in Questions 6 and 7.

LAB 38
Synoptic Weather Maps

Purpose

The purpose of this lab is to have you read and interpret the information displayed on synoptic weather maps. You will also learn the techniques used by meteorologists to analyze weather maps and the weather conditions associated with frontal systems.

Materials

colored pencils
copies of the following surface weather maps available from
   <http://www.ametsoc.org/dstreme/>
current isobar, front, radar, and data surface chart
current 500 mb-data upper air chart

Procedure A
Complete the following steps.

1. On a separate piece of paper, decode and write out the current weather conditions for the ten station models in Figure 38–1.

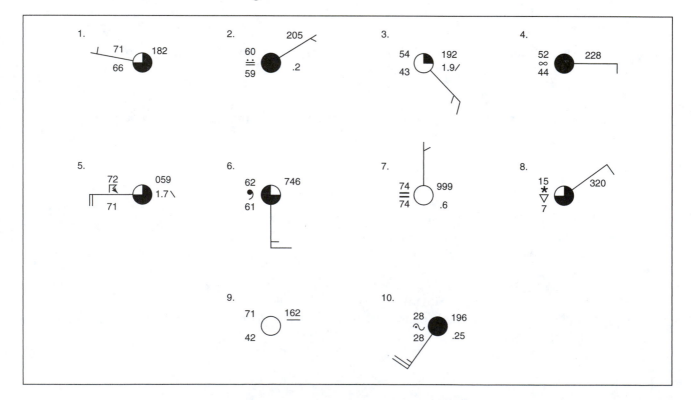

FIGURE 38–1   Station Models

2. On the back of the same piece of paper you used to decode the ten station models, encode the following six current weather conditions in station model format.

Station Model A—The sky is clear. The wind is from the south at five knots. The barometric pressure is 1024.8 mb and it has risen 2.4 mb in the past three hours. The air temperature is 74 degrees F and the dew point is 66 degrees F.

Station Model B—The wind is from the northwest at 15 knots. Cloud cover is 25 percent. The present weather is drizzle. There has been .2 inches of rain in the past six hours. The barometric pressure is 974.6 mb and it has fallen 9.2 mb in the last three hours. The air temperature is 62 degrees F and the dew point is 61 degrees F.

Station Model C—The sky is totally overcast. The present weather is fog. The air temperature is 76 degrees F and the dew point is 74 degrees F. There has been .6 inches of rain in the past six hours. Barometric pressure is 999.9 mb and has fallen 3.6 mb in the past three hours. The wind is from the north at 20 knots.

Station Model D—The sky is 75 percent covered. The wind is from the southeast at 25 knots. The barometric pressure is 999.0 mb and has dropped 7.2 mb in the past three hours. It is raining and the air temperature is 79 degrees F, and the dew-point is 74 degrees F. There has been .5 inches accumulation of rain in the past six hours.

Station Model E—The sky is clear. The wind is calm. The barometric pressure is 983.0 mb and rising. The air temperature is 70 degrees F and the dew point temperature is 62 degrees F. The visibility is fi mile with no precipitation.

Station Model F—The sky is half covered. Thunderstorms are threatening. There has been .1 inches accumulation of rain. There is a westerly wind at 10 knots. The barometric pressure is 1005.8 mb and has dropped 8.2 mb in the past three hours. The air temperature is 72 degrees F and the dew point is 71 degrees F.

Procedure B

Complete the following steps.

1. Use the current surface weather map data from the Data Streme Web site to draw the current weather conditions on the blank map in Figure 38–2 with the appropriate colors. Your current weather map should include the locations of the high- and low-pressure centers, fronts, and precipitation.
2. With help from your instructor, use the current 500-mb upper air chart from the Data Streme Web site to locate the jet stream. Upper air maps can be analyzed by drawing an arrow through each wind vane of the individual weather station pointing in the direction the wind is moving. The jet stream can be located by identifying the band of the highest wind velocities. Once you have identified the jet stream's location, draw it in using a green colored pencil on your current weather map.
3. Using the data on your current weather map, the location of the jet stream, and help from your instructor, predict what the weather conditions will be in 24 hours. On the blank map in Figure 38–2, draw the locations of the high and low pressure, fronts, and precipitation to make a forecast map for the next day.

Current weather

Tomorrow's forecast

FIGURE 38–2   Weather Forecasts

Procedure C

Using your textbook, fill in the appropriate weather conditions associated with each type of front on Table 38–1.

TABLE 38–1 Fronts		
	Cold	Warm
Air Temp. in Front		
Air Temp. Behind		
Cross-Section		
Cloud Types		
Pressure in Front		
Pressure Behind		
Frontal Symbol		
Precipitation		

147

Conclusions

1. Describe the change in wind direction associated with a low-pressure system moving through your area.

2. What general weather conditions are associated with high-pressure systems?

3. What would you expect the weather conditions to be like if the barometric pressure trend over the past three hours has shown a steady decrease?

4. What is the jet stream, and why is it an important aspect of weather forecasting?

5. Explain what information you need to make accurate weather forecasts.

LAB 39
Clues to Past Climate

Purpose

The purpose of this lab is to have you use similar scientific techniques that paleoclimatologists use to infer about the Earth's past climate.

Materials

metric ruler	graph paper or
colored pencils	computer spreadsheet application

Procedure A

Using the diagram of a core sample from a Norwegian Spruce tree in Figure 39–1, measure the growth in millimeters for each ring of seasonal growth. Record your measurements in Table 39–1.

Next, use the data in Table 39–1 to create a dual-line graph that shows the relationship between tree growth and annual precipitation. The x-axis should be labeled "Year," the first y-axis should be labeled "Tree Ring Growth (mm)," and the secondary y-axis should be labeled "Average Precipitation (inches)." Use a different color for each line, and label them accordingly.

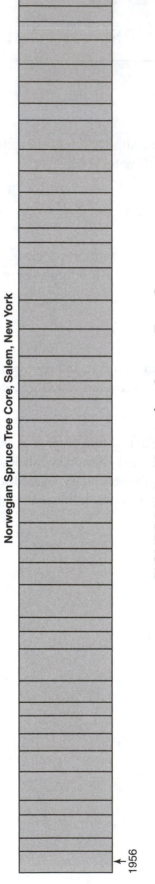

Norwegian Spruce Tree Core, Salem, New York

1997

1956

FIGURE 39–1   Norwegian Spruce Tree Core

TABLE 39-1	Tree Growth/Average Precipitation	
Year	Tree Ring Growth (mm)	Average Precipitation (inches)*
1956		5
1957		2.8
1958		4.3
1959		2.3
1960		7.4
1961		2.4
1962		2.8
1963		3.1
1964		0.9
1965		4.1
1966		6
1967		1.9
1968		2.8
1969		3
1970		3.8
1971		3.4
1972		1.5
1973		2.5
1974		6.3
1975		6.9
1976		3.8
1977		7.4
1978		2.5
1979		5.7
1980		2.2
1981		4.1
1982		2.3
1983		2.6
1984		1.5
1985		4.5
1986		2
1987		7.3
1988		2.2
1989		5
1990		1.7
1991		4.5
1992		3.2
1993		4.9
1994		3.5
1995		3
1996		5.8
1997		3.3

*Average precipitation data for the Hudson Valley Region, New York State.

Procedure B

Research teams have used data gathered from ice cores taken in Antarctica to determine the Earth's past atmospheric carbon dioxide concentration and change in global temperature. Use the ice core data in Table 39–2 to create a dual-line graph showing the relationship between atmospheric carbon dioxide concentration and global temperature. The x-axis should be labeled "Thousands of Years Ago," the first y-axis should be labeled "Carbon Dioxide Concentration (ppt), and the second y-axis should be labeled "Change in Global Temperature (°C)." Use a different color for each line and label them accordingly.

TABLE 39–2 Antarctic Ice Core Data		
Thousands of Years Ago	Carbon Dioxide	Change in Global Temperature
160	1.95	–9
150	2.05	–9.5
140	2.3	–7.5
130	2.95	–2
120	2.8	–2.5
110	2.7	–7
100	2.4	–4
90	2.4	–6.5
80	2.3	–5
70	2.4	–6.5
60	1.95	–7.7
50	2.18	–7.5
40	1.9	–7
30	2.2	–9
20	1.95	–10
10	2.55	–0.5
0	3.60	2.5

Conclusions

1. Using your graph that compares seasonal tree ring growth to average annual precipitation, describe the relationship between tree growth and precipitation.

2. Describe how the relationship you identified between tree ring growth and average precipitation can be used to infer past climate.

3. Using your graph of ice core data, describe the relationship between atmospheric carbon dioxide concentration and global temperature.

4. According to your ice core data, what happens to global temperature when carbon dioxide decreases?

5. According to your ice core data, what happens to global temperature when carbon dioxide increases?

6. According to your data, how long ago did the last ice age begin?

7. According to your data, how long ago did the last ice age end?

LAB 40
Humans and Global Climate Change

Purpose

In the following lab, you will explore the influences that human beings and our technology have on global climate. Topics such as acid precipitation, photochemical smog, and global warming are all examples of how humans have the ability to affect the environment in such a way as to alter local, regional, and global climate. The results of these actions are unprecedented changes occurring in the Earth's climate in a relatively short period of time.

Materials

sample of coal   balance or scale

Procedure A—Sulfur Dioxide Emissions

In this activity, you will predict the amount of sulfur dioxide (SO_2) produced as a result of burning coal. Use the following information to help you with your calculations: coal contains approximately 1.1 percent sulfur.

1. Using the piece of coal provided, determine its mass to the nearest tenth of a gram, then calculate the amount of sulfur dioxide emitted into the atmosphere by completely burning your sample. Assume that all of the sulfur contained within the coal (1.1 percent) is converted into sulfur dioxide. This is calculated by multiplying the mass of your coal sample by .011. Record your answer below. Show your work.

2. Coal consumption in the United States during 1995 amounted to 864 million tons. Using this data, determine the amount of sulfur dioxide put into the atmosphere by burning coal during 1995. This can be determined by multiplying the U.S. annual coal consumption by 0.011. Record your answer below. Show your work.

3. The U.S. Environmental Protection Agency (EPA) is attempting to lower the amount of acid-forming compounds put into the atmosphere by controlling fossil fuel emissions. Their proposed target emissions for sulfur dioxide in 1995 was seven million tons. According to your sulfur dioxide emission calculations from the previous question, did the EPA meet its goals?

Procedure B—Carbon Dioxide and Global Warming

In this activity, you will determine the effects of automobile emissions on the amount of carbon dioxide in the atmosphere.

1. Determine, to the best of your ability, how many gallons of gasoline are used by you or your family's motor vehicles each week. Assume that the average car uses approximately 15 gallons per week. Record your answer below. Show your work.

2. Next, determine the amount of gasoline used by your motor vehicle annually. This can be determined by multiplying your weekly gasoline use from the previous question by 52 weeks per year. Record your answer below. Show your work.

3. Use the following information to determine the amount of carbon dioxide (CO_2) generated by your family's vehicles and emitted into the atmosphere annually: one gallon of gasoline when burned emits 21.7 pounds of carbon dioxide into the atmosphere. Multiply your annual use of gasoline by 21.7, and record your answer below. Show your work.

4. Scientists have determined that, on average, a young growing tree removes 25 pounds of carbon dioxide from the atmosphere each year. Use your annual carbon dioxide emissions from Question 3 to determine how many trees you or your family needs to remove all of the carbon dioxide they produce. This is calculated by dividing your total annual carbon dioxide production by 25. Record your answer below. Show your work.

5. An average healthy forest contains approximately 400 trees per acre. Knowing this, how many acres of forest would you or your family require to remove all of the carbon dioxide generated by your family's motor vehicles? To calculate this, you must divide the number of trees required to remove the carbon dioxide you determined in Question 4 by 400. Record your answer below. Show your work.

Conclusions

1. Based on your calculations from Procedure A, do you think it is possible that the burning of coal has led to an increase in the amount of sulfur dioxide in the atmosphere?

2. Explain the negative effects of having increased amounts of sulfur dioxide in the atmosphere.

3. Based on your calculations from Procedure B, do you think it is possible that humans have increased the amount of carbon dioxide in the atmosphere?

4. Do you and your family have enough trees on your property to remove all of the carbon dioxide you produce?

5. If half of all Americans (approximately 160 million people) produce the same amount of carbon dioxide as your family, are there enough forests in the country to remove it all? Assume that the entire forest land of the United States is 736,681,000 acres. Show your work.

6. Now take your answer from Question 5 and double it to account for all of the commercial vehicles that produce carbon dioxide such as trucks, buses, and planes. Is there enough forest land in the United States to remove it all?

LAB 41
The Effects of Acid Precipitation on Building Materials

Purpose

The purpose of this lab is to have you observe and record the effects of acid precipitation on common building materials.

Materials

500 ml of dilute acid solution with a
   pH of 3.0 and 4.0
five 250-ml beakers
tongs
balance or scale
safety glasses
colored pencils

pH test kit
pieces of brick, concrete, iron, marble, and
   painted metal
paper towels
heat lamps
graph paper or computer spreadsheet
   application

SAFETY CONCERN
THE USE OF WEAK HYDROCHLORIC ACID SOLUTION TO SIMULATE ACID PRECIPITATION SHOULD BE CONTROLLED BY THE INSTRUCTOR, AND SAFETY GLASSES SHOULD BE USED AT ALL TIMES WHEN WORKING WITH THE ACID SOLUTION!

Procedure

Complete the following steps.

1. Obtain samples of the five building materials provided by your instructor. Label five 250-ml beakers with letters A through E.
2. Assign each building material sample a lettered beaker, and record the name of each specific material in the correct space in Table 41–1.
3. Using a balance or scale, determine the mass of each sample to the nearest tenth of a gram. Record each building materials start mass in Table 41–1.
4. Place each sample in the correctly labeled beaker and obtain the dilute acid solution from your instructor. While wearing your safety glasses, carefully pour enough acid solution into each beaker to completely cover each sample of building material.
5. This long-term experiment will determine the possible effects of acid rain on certain building materials. Each day in class, carefully remove each building material sample from the beaker using tongs, blot it dry with a paper towel, and place it under the heat

lamp set up by your instructor. It is essential that each sample be dried thoroughly before weighing it. Weigh each sample to the nearest tenth of a gram and return it carefully to the correct beaker. Record this mass in Table 41–1. Repeat this procedure each day for the length of the experiment.

6. After your experiment has been completed, follow your instructor's procedures for proper cleanup.

7. Using the data gathered from your experiment, create a multiple-line graph that shows the change in the mass of your building materials over time. The x-axis should be labeled "Time in Days" and the y-axis should be labeled "Building Material." Make sure you use a different colored pencil for each building material plotted on your graph. Make a key and label each line on your graph.

TABLE 41–1	Mass of Building Materials							
Beaker	Building Material	Mass Day 1	Mass Day 2	Mass Day 3	Mass Day 4	Mass Day 5	Mass Day 6	Mass Day 7
A								
B								
C								
D								
E								

Conclusions

1. Calculate the rate of deterioration for each building material by using the following formula and show your work:

$$\text{change in mass} \div \text{change in time}$$

   A. Brick

   B. Concrete

   C. Iron

   D. Marble

   E. Painted metal

2. Which building material was most resistant to the effects of acid precipitation? How does your data support this conclusion?

3. Which building material was least resistant to the effects of acid precipitation? How does your data support this conclusion?

4. Explain the effects the acid solution had on the painted metal.

5. Briefly describe how acid precipitation forms in the atmosphere.

6. Using the results of your experiment, discuss which structures would be most negatively affected by acid precipitation.

LAB 42
Effects of Acid Precipitation on Seed Germination

Purpose

The purpose of this lab is for you to determine the effects of acid precipitation on the germination of seeds. Determining the germination rate of seeds exposed to acid precipitation will reveal the effects of secondary air pollution on agriculture and food production.

Materials

seeds
paper towels
plastic bags
wash bottles
rubber gloves

acid precipitation solution (pH between 3.0 and 4.0)
tape
pH test kits

SAFETY CONCERN

THE USE OF WEAK ACID SOLUTION TO SIMULATE ACID PRECIPITATION SHOULD BE CONTROLLED BY THE INSTRUCTOR, AND SAFETY GLASSES AND RUBBER GLOVES SHOULD BE USED AT ALL TIMES WHEN WORKING WITH THE ACID SOLUTION!

Procedure

Complete the following steps.

1. Place ten seeds in a paper towel and roll it up around the seeds. Use tape to keep it folded. While wearing rubber gloves and safety glasses, soak the paper towel with the acid precipitation solution, and place the wet paper towel in a plastic bag. Seal the bag closed with tape, and label the bag with your group's name and "Acid Rain."
2. Repeat step one, but this time, soak the paper towel in tap water. Place the paper towel in another plastic bag. Seal the bag closed with tape, and label the bag with your group's name and "Tap Water." Place both bags in a warm location.
3. Use a pH test kit to determine the pH of both the acid precipitation and the tap water. Record the values in Table 42–1.
4. Open both bags after about five to seven days and count the seeds that have germinated. Determine the percent germination rate by dividing the number of seeds germinated by ten. Record the results in Table 42–1. Show your work.

TABLE 42–1 pH Values and Germination Percentages		
Sample	pH Value	Percent Germination

Conclusions

1. What was the percent of germination for the seeds that were exposed to tap water?

2. What was the percent of germination for the seeds exposed to acid precipitation?

3. Describe the effects of acid precipitation on the germination rate of seeds.

4. Describe the possible effects that reduced seed germination would have on agriculture.

LAB 43
Effects of Acid Precipitation on Plant Growth

Purpose

The purpose of this lab is to determine the effects of acid precipitation on living plants. Determining the effect of plants exposed to acid precipitation will reveal the effects of secondary air pollution on agriculture and food production.

Materials

plant seedlings (bean seedlings are good
  because they grow large very quickly)
pots
tape
pH test kits

acid precipitation solution (pH between 3.0
  and 4.0)
potting soil
wash bottles
plastic trays

SAFETY CONCERN
**THE USE OF WEAK ACID SOLUTION TO SIMULATE ACID PRECIPITATION
SHOULD BE CONTROLLED BY THE INSTRUCTOR, AND SAFETY GLASSES
SHOULD BE USED AT ALL TIMES WHEN WORKING WITH THE ACID SOLUTION!**

Procedure

Complete the following steps.

1. Transplant two plant seedlings into two pots provided by your instructor. Label one pot "Acid Precipitation" and the other "Tap Water." Also label each pot with the name of your group.
2. Place each plant in the appropriate tray provided by your instructor. One tray will be labeled "Acid Rain" and the other "Tap Water."
3. Irrigate the "Acid Rain" plants with the acid precipitation solution in the wash bottle. Make sure to soak the foliage of the plant as well as the soil. Repeat this procedure on the "Tap Water" plants using the wash bottle filled with tap water.
4. Use a pH test kit to determine the pH of both the acid precipitation and the tap water. Record the values in Table 43–1.
5. Over a period of one to two weeks, continue to irrigate and wash each plant each day with the appropriate water, and observe and record the effects on each plant. Record your observations in Table 43–1.

TABLE 43–1 Tap Water and Acid Rain pH Levels			
Tap Water pH =		Acid Rain pH =	
Date	Observation	Date	Observation

Conclusions

1. What was the pH value you determined for the simulated "Acid Rain"?

2. What was the pH value you determined for the "Tap Water"?

3. Describe the overall appearance of the plant bathed in "Acid Rain."

4. Explain why this experiment may not be a true representation of the effects of acid precipitation.

5. Describe the possible effects that acid precipitation may have on agriculture.

LAB 44
Stratospheric Ozone Depletion

Purpose

The purpose of this lab is to have you identify the depletion of ozone gas in the stratosphere over a 40-year period and its relationship to increased amounts of deadly ultraviolet radiation reaching the Earth's surface.

Materials

graph paper or
   computer spreadsheet program
colored pencils

Procedure

Complete the following steps.

1. Using the data on stratospheric ozone concentration shown in Table 44–1, create a line graph that plots the concentration of ozone gas for each year. Label the x-axis "Year" and the y-axis "Total Ozone (Dobson Units)."

TABLE 44–1 Stratospheric Ozone Concentration	
Year	Total Ozone (Dobson Units)
1956	324
1957	327
1958	312
1959	310
1960	300
1961	315
1962	330
1963	311
1964	315
1965	280
1966	313
1967	320
1968	300
1969	280
1970	280
1971	295
1972	305
1973	285
1974	275
1975	305
1976	280
1977	250
1978	282
1979	255
1980	225
1981	230
1982	227
1983	210
1984	200
1985	195
1986	250
1987	160
1988	225
1989	165
1990	175
1991	150
1992	145
1993	120
1994	125

2. Using the data on the relationship between ozone loss and ultraviolet radiation striking the Earth's surface in Table 44–2, create a line graph that shows this relationship. Label the *x*-axis "Percent Ozone Change" and the *y*-axis "Percent Change in UV Radiation at the Earth's Surface."

TABLE 44–2 Ozone Loss and Ultraviolet Radiation	
Percent Ozone Change	Percent Change in UV Radiation at the Earth's Surface
0	0
5	5
10	12
15	20
20	28
25	36
30	47
35	60
40	76
45	92
50	116
55	130
60	150

Conclusions

1. Does your data support the theory that ozone gas in the stratosphere is being depleted?

2. Calculate the rate of ozone lost in the stratosphere between the years 1956 and 1994 by using the following formula and show your work:

$$\frac{\text{change in ozone}}{\text{change in time}}$$

3. Describe the relationship between loss of stratospheric ozone and ultraviolet radiation at the surface.

4. Calculate the percentage loss of stratospheric ozone between 1956 and 1994. Show your work.

5. Using the answer from the previous question, determine what percentage the ultraviolet radiation reaching the Earth's surface increased for the same time period.

6. Explain the importance of ozone gas in the stratosphere.

7. Describe the negative effects of increasing amounts of ultraviolet radiation reaching the Earth's surface.

LAB 45
The World's Oceans

Purpose

The purpose of this lab is for you to identify the main characteristics of the world's oceans including their size, major currents, and relationship between temperature and depth.

Materials

graph paper or
  computer spreadsheet application
colored pencils

Procedure A

Using your textbook and the blank map of the world (Figure 45–1), draw and label all of the major ocean currents. Use a blue colored pencil to draw the cold currents and a red colored pencil for the warm currents.

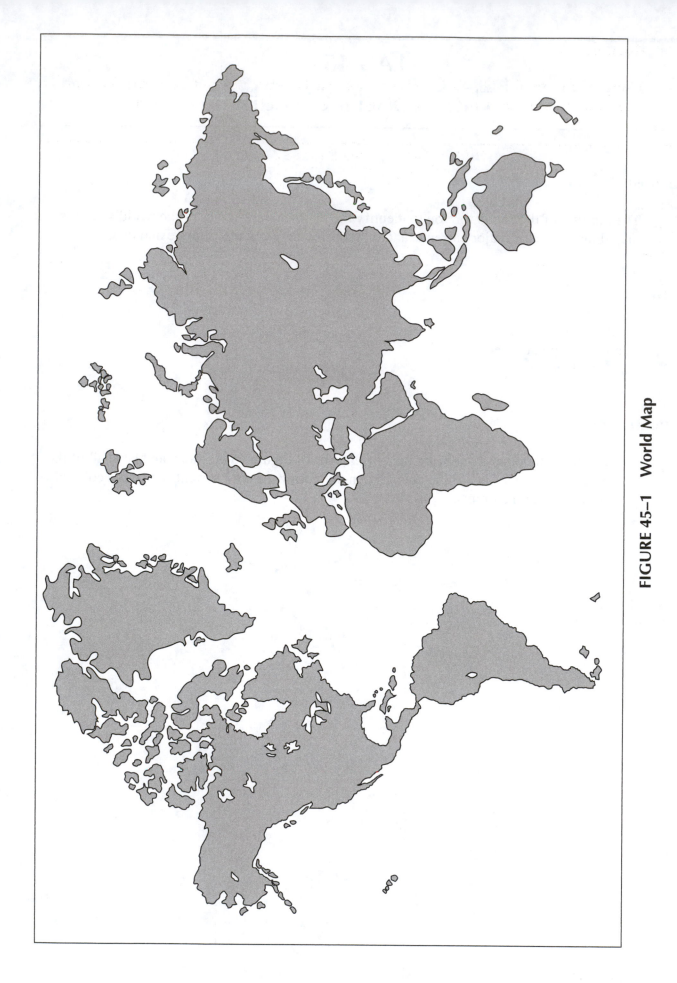

FIGURE 45–1   World Map

Procedure B

Using the data from Table 45–1, create a bar graph that shows the surface area for the world's three main oceans. Label the x-axis "Ocean" and the y-axis "Millions of Kilometers."

TABLE 45–1 Ocean Surface Areas	
Ocean	Surface Area (million km)
Atlantic	106.6
Pacific	181.3
Indian	74.1

Procedure C

Use the data from Table 45–2 to create a line graph that shows the relationship between the temperature and depth within the ocean near the tropics. Label the x-axis "Temperature (°C)" and the y-axis "Depth (meters)."

TABLE 45–2 Ocean Temperatures and Depth	
Temperature (°C)	Depth (meters)
16.0	0
15.5	250
14.0	500
8.0	750
5.0	1000
3.5	1250
1.5	1500
1.0	1750
.5	2000
.25	2250
0	2500
0	3000
0	3500
0	4000

Conclusions

1. Which ocean currents bring cool, nutrient-rich water from the poles toward the equator?

2. Which ocean currents distribute warm water from the equator?

3. Calculate the percentage that each of the three major oceans occupies out of the total ocean area on the Earth.

4. The atmospheric pressure directly above the ocean surface is approximately 14.7 pounds per square inch, and for every 33 feet of depth, the pressure increases another 14.7 pounds per square inch. Knowing this, calculate the pressure for the average depth of the world's oceans (12,171 feet). Show your work.

5. List the world's oceans from largest to smallest.

6. Define the term *thermocline*. At what depths does it occur within the tropical oceans?

LAB 46
Discharge Rate Investigation

Purpose

The purpose of this lab is for you to learn how the discharge rate can be determined for a flowing body of water.

Materials

graph paper

Procedure

The discharge rate measures how much water is moving past a particular point in a river or stream. It is calculated by using the following formula:

Q = AV
Q = discharge rate
A = cross-sectional area of river or stream
V = velocity of river or stream

1. Using the data from Table 46–1, create a line graph that shows the cross-section of a portion of the White Creek, which is a tributary of the Battenkill and Hudson Rivers in New York State. The x-axis will be labeled "Distance From Left Bank (feet)" and the y-axis will be labeled "Depth (inches)." Because your y-axis is the depth, number up the side of your graph starting with the greatest depth (11 inches) and count backwards to zero, which represents the surface of the water. Plot each data point and connect them with a line.

2. All of the squares within your graph represent 12 square inches (1 inch of depth × 12 inches of width per square). Therefore, determining the number of squares within your cross-section and multiplying it by 12 will give you the cross-sectional area in inches. Record your answers on Table 46–2.

3. Next, convert your cross-sectional area into square feet by dividing your answer from Question 2 by 144 (144 square inches per square foot). Record your answer in Table 46–2.

4. The velocity of the White Creek was determined by throwing a ball in the creek and recording the time it took to travel 10 feet. The average velocity was determined to be 0.43 feet per second. Now, calculate the discharge rate for the White Creek. Record your answer in Table 46–2.

5. Using the equivalent of 7.5 gallons per cubic foot, convert the discharge rate for the White Creek into gallons per second. Record your answer in Table 46–3. Complete Table 46–3 for discharge rates shown.

TABLE 46–1 Distance and Depth Table

Distance from Left Bank (ft)	Depth (inches)
0	0
1	6
2	9
3	9
4	10
5	11
6	10
7	10
8	7
9	7
10	5
11	5
12	5
13	3
14	2
15	0

TABLE 46–2 Velocity Table

Number of Squares	Cross-Sectional Area (in²)	Cross-Sectional Area (ft²)	Velocity (ft/sec)	Discharge Rate (ft³/sec)

TABLE 46–3 Discharge Rate Table

Discharge Rate (ft³/sec)	Discharge Rate (gal/sec)	Discharge Rate (gal/min)	Discharge Rate (gal/hour)	Discharge Rate (gal/day)	Discharge Rate (gal/year)

Conclusions

1. What is the source of all the water that is flowing in rivers and streams?

2. During which season do you believe a river or stream to have the greatest discharge rate, and why?

3. During which season do you expect the lowest discharge rates to occur, and what is the main source of water in streams and rivers during this time of year?

4. The White Creek flows into a larger river called the Battenkill. Calculate the discharge rate for the Battenkill River if its cross-sectional area was determined to be 139.5 feet and its velocity 3.3 feet per second.

LAB 47
Fresh Water Testing Lab

Purpose

The purpose of this lab is to have you analyze water samples of local freshwater resources to identify their relative health.

Materials

nitrate nitrogen test kit   wide-range pH test kit
ortho phosphate test kit   water samples

Procedure:

Using the water samples provided, test for the following chemical composition.
Nitrate (NO_3^-)
   Nitrate is a form of nitrogen that is an essential nutrient for plants
   and animals, and is used to build proteins. Nitrate is measured in parts per million (ppm)
   and milligrams per liter (mg/L). Nitrate is found naturally in water as a result of plant
   growth and decay, especially manure waste.

Nitrate Levels
   Typical natural levels for fresh water—less than 1 ppm
   Recommended level for trout—less than .06 ppm
   Human health risks and drinking water standards—less than 20 ppm
   Sewage treatment plant effluent—30 ppm

Nitrate Test Results

pH
   pH is a measure of the acidity or alkalinity of a solution, and it can affect the biological
   and chemical properties of water. Levels from 0 to 7 are considered acidic, levels between
   7 and 14 are considered alkaline, and 7 is neutral.

pH Levels
   Optimal range for most life—6.5 to 8.2
   Federal drinking water standards—6.5 to 8.5

pH Test Results

Phosphate (PO_4^{-3})

Phosphate is a plant nutrient found in phosphate containing rocks, oil, and animal waste. Phosphate can also be found in detergents. Excess phosphate can greatly affect aquatic ecosystems by causing a rapid growth of aquatic plants that disrupt the aquatic ecosystem.

Phosphate Levels

Natural levels—less than 0.3 ppm
Above 0.1—impact likely
Above 0.5—impact certain
Federal drinking water standards—less than 0.5 ppm
Wastewater effluent—5 to 30 ppm

Phosphate Test Results

Conclusions

1. What are the sources of nitrate in freshwater bodies?

2. At what levels does nitrate begin to affect the health of humans if found in their drinking water?

3. At what levels are nitrates found in unpolluted water?

4. Describe the amount of nitrate found in your water sample and how it would affect living things.

5. What are the sources of phosphate in freshwater bodies?

6. At what levels do phosphates begin to affect the health of humans if found in their drinking water?

7. At what levels are phosphates found in unpolluted water?

8. Describe the amount of phosphate found in your water sample and how it would affect living things.

9. What is the optimum pH for a healthy body of fresh water?

10. Describe the pH levels that were determined in your water samples.

11. Describe the overall quality of your water sample.

LAB 48
Dissolved Oxygen and Biological Oxygen Demand

Purpose

The purpose of this lab is to have you identify the relationship between the amount of dissolved oxygen in the water and its ability to support aquatic life. You will also reveal how water pollutants can affect the amount of dissolved oxygen in the water.

Materials

graph paper or computer spreadsheet application

Procedure A

Using the data from Table 48–1, create a line graph that shows the relationship between dissolved oxygen and water temperature. The x-axis will be labeled "Temperature (°F)" and the y-axis will be labeled "Dissolved Oxygen (ppm)."

TABLE 48–1 Temperature and Dissolved Oxygen

Temperature (°F)	Maximum Dissolved Oxygen (ppm)
32	14.6
33.8	14.2
35.6	13.8
37.4	13.5
39.2	13.1
41	12.8
42.8	12.5
44.6	12.2
46.4	11.9
48.2	11.6
50	11.3
51.8	11.1
53.6	10.8
55.4	10.6
57.2	10.4
59	10.2
60.8	10
62.6	9.7
64.4	9.5
66.2	9.4
68	9.2
69.8	9
71.6	8.8
73.4	8.7
75.2	8.5
77	8.4
78.8	8.2
80.6	8.1
82.4	7.9
84.2	7.8
86	7.6
87.8	7.5
89.6	7.4
91.4	7.3
93.2	7.2
95	7.1
96.8	7
98.6	6.8
100.4	6.7
102.2	6.6
104	6.5

Procedure B

Biological oxygen demand (BOD) is the amount of oxygen that is required by aerobic bacteria to break down organic material in an aquatic ecosystem. Using the data from Table 48–2, create a dual-line graph that shows the dissolved oxygen and biological oxygen demand for a stretch of the Fishkill River. The x-axis will be labeled "Thousands of Feet Down River," the primary y-axis will be labeled "Dissolved Oxygen (ppm)," and the secondary y-axis will be labeled "Biological Oxygen Demand (ppm)."

	A	B	C
1	Thousands of Feet	DO Level (ppm)	BOD Level (ppm)
2	1	8	0.5
3	2	8	0.5
4	3	8	0.5
5	4	8	0.5
6	5	8	0.5
7	6	8	0.5
8	7	8	0.5
9	8	7.5	8.5
10	9	7	8.5
11	10	6.5	8.25
12	11	5.75	8.25
13	12	4	8
14	13	2.5	7.75
15	14	1.5	7.5
16	15	1	6.5
17	16	0.5	6.25
18	17	0.5	5.5
19	18	0.75	5
20	19	1.5	4.5
21	20	2.25	3.75
22	21	3	3
23	22	4	2.5
24	23	4.5	2
25	24	5.5	1.75
26	25	6	1.5
27	26	6.25	1.25
28	27	7.5	1
29	28	7.75	0.75
30	29	8	0.5
31	30	8	0.5
32	31	8	0.5
33	32	8	0.5

TABLE 48–2 DO and BOD Levels of the Fishkill River

Conclusions

1. Using your graph from Procedure A, at what temperature is the dissolved oxygen concentration the highest? What is the level of oxygen at this temperature?

2. Using your graph from Procedure A, at what temperature is the dissolved oxygen concentration the lowest? What is the level of oxygen at this temperature?

3. If the average temperature of a river is 52 degrees F, what is the highest possible dissolved oxygen level it could have?

4. What is the normal dissolved oxygen level for the Fishkill River?

5. What is the normal biological oxygen demand for the Fishkill River?

6. At what location is the pollution entering into the Fishkill River?

7. What effect does the pollution have on the river's dissolved oxygen level?

8. What effect does the pollution have on the river's biological oxygen demand?

9. What is the relationship between dissolved oxygen levels and biological oxygen demand in the Fishkill River?

10. How many feet of river is affected by the pollution? How many feet of river would experience fish kill (levels below 4 ppm)?

11. At what distance would the river return to normal conditions?

LAB 49
Biomes of the World

Purpose

The purpose of this lab is for you to identify the locations, climate characteristics, and productivity associated with the major biomes of the world.

Materials

graph paper or computer spreadsheet application
colored pencils

Procedure A

Using your textbook as a reference and colored pencils, shade in the different locations of the nine major world biomes on the blank map in Figure 49–1. Use a different color for each biome type and create a key.

FIGURE 49–1　World Biomes Map

Procedure B

Using the data from Table 49–1, create a combination line-bar graph that shows the relationship between a biome's mean annual temperature and mean annual precipitation. Plot the data points and connect them with a line for the mean annual temperature. Use bars to represent the mean annual precipitation. The x-axis will be labeled "Biomes," the primary y-axis will be labeled "Mean Annual Temperature (°F)," and the secondary y-axis will be labeled "Mean Annual Precipitation (inches)."

TABLE 49–1	Biomes and Mean Annual Temperatures and Precipitation	
Biome	Mean Annual Temperature (°F)	Mean Annual Precipitation (inches)
Desert	69	8
Savannah	80	50
Grassland	61	20
Deciduous Forest	67	40
Coniferous Forest	45	55
Tundra	26	22
Tropical Rainforest	80	120

Procedure C

Using the data from Table 49–2, create a bar graph that shows the relationship between the world's ecosystems and their productivity. The *x*-axis will be labeled "Ecosystems" and the *y*-axis will be labeled "Annual Productivity (kilograms per square meter)."

	A	B
	TABLE 49–2 Ecosystem Productivity (kg/m²/year)	
1	Ecosystem	Productivity
2	Estuaries	8800
3	Swamps and Marshes	8800
4	Tropical Rainforest	8800
5	Deciduous Forest	5800
6	Coniferous Forest	3500
7	Savannah	3000
8	Agricultural Land	2600
9	Grassland	2200
10	Lakes and Streams	2200
11	Continental Shelf (Coastal Ocean)	1500
12	Open Ocean	1200
13	Tundra	600
14	Desert	350

Conclusions

1. Describe the relationship between latitude and temperature for the world's biomes.

2. What is the relationship between the temperature, precipitation, and productivity in an ecosystem?

3. Describe the annual precipitation and temperature values of two specific ecosystems and how their productivity is affected by them.

4. Which two terrestrial ecosystems have the lowest and the highest productivity?

5. Which two aquatic ecosystems have the lowest and the highest productivity?

6. Where would you most likely catch the most fish: the open ocean or the coastal ocean? Why?

7. Which two ecosystems have similar productivity as that of agricultural land?

LAB 50
Biotic and Abiotic Components
of Local Ecosystems

Purpose

The purpose of this lab is to identify the biotic and abiotic elements of two local ecosystems, one disturbed and one undisturbed. Your instructor will take you to two different outdoor locations, where you will perform a survey of all the biotic and abiotic elements that make up each ecosystem. The first ecosystem you will survey will represent an undisturbed ecosystem. An undisturbed ecosystem is one that has not been exposed to intense human activity. The second survey will investigate a disturbed ecosystem that has been exposed to intense human activity. Make sure that both the undisturbed and the disturbed ecosystems are similar ecological systems. For example, if you have a forest nearby, choose a pristine section of the forest to represent the undisturbed ecosystem. Choose another section of the forest that has been exposed to human activity to represent the disturbed ecosystem. Forest ecosystems work best for this lab, although other ecosystem types may also be used.

Materials

measuring tape
vinyl ribbon tape
field guides
thermometer
humidity sensor
compass
clinometer
soil pH meter or test kit
light meter
small paper bags
2-liter plastic soda bottle
watch glass
isopropyl alcohol
forceps
coarse screen
ring stand
heat lamp
dissecting microscope or hand lens

Procedure A

1. For each ecosystem, first identify the species of organisms you find within a 10-square meter plot. Have your instructor help you identify a good place for your survey. Once you have set up your perimeter, mark it with your vinyl ribbon tape. Use your field guides or ask your instructor to help you identify the biotic components within your perimeter. Make a list of all the species you have identified.

2. Next, count the number of each individual species you have identified within your study area. Try to be as accurate in your counting as possible.

3. After you have completed your biotic survey, identify and record some of the abiotic components in your study area. These will include the following: air temperature at 1 meter above the ground, soil temperature at a depth of 2 inches, humidity 1 meter above the ground, light level 1 meter above the ground, soil pH, slope of the study area, compass direction the study area faces, color of soil, and rock types if available.

4. Finally, take a sample of the leaf litter at the ground surface. Scoop up a large handful of organic debris at the soil surface, and place it in your paper bag. Make sure to label it so that you can tell the difference between the disturbed and undisturbed leaf litters.

5. Repeat the above procedures for both the disturbed and undisturbed ecosystems. It may take more than one day to complete both site surveys.

6. After your surveys are complete, return to the lab with your bags of leaf litter, and with help from your instructor, set up two Berlese-Tullgren funnels (see Figure 50–1), one for each ecosystem. This will help you identify the organisms that live within the leaf litter at the ground surface. Place your leaf litter into your funnel and turn on the lamp. The heat from the lamp will cause any organisms within the leaf litter to move down away from the light and heat, and they will then fall into the watch glass filled with alcohol. Leave the funnel set up for 24 hours.

SAFETY CONCERN
MAKE SURE THAT THE LIGHT IS NOT TOO CLOSE TO THE PLASTIC BOTTLE OR LEAF LITTER—THIS MAY RISK MELTING OR BURNING THE MATERIAL!

7. The next day, turn off the light and examine the watch glass for organisms using a dissecting microscope or hand lens. Identify the organisms present in the leaf litter and add them to your inventories. You may choose to just identify the organisms as either an insect, arthropod, or worm, and how many of each there are.

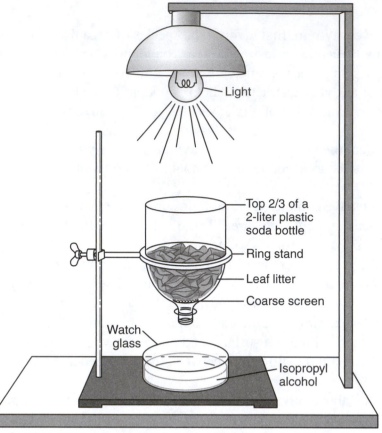

Light

Top 2/3 of a
2-liter plastic
soda bottle

Ring stand

Leaf litter

Coarse screen

Watch
glass

Isopropyl
alcohol

Caution: Make sure the light is not too close to the leaf litter or the plastic bottle, this may risk melting or burning of the material!

FIGURE 50–1

Procedure B

1. For each ecosystem, calculate the biotic abundance for each species of organism you identified in your survey using the following formula:

$$\text{biotic abundance (\%)} = \frac{\text{number of species}}{\text{total \# of all organism}} \times 100$$

For example, if you counted 10 white pine trees within your survey, and the total number of organisms within your plot was 100, the biotic abundance of white pine trees would be calculated as 10 percent.

2. Using the biotic abundance data for each ecosystem, create two bar graphs that shows the abundance for the disturbed and undisturbed ecosystems. The x-axis should be labeled "Organism," and the y-axis should be labeled "Percent Abundance."

3. Prepare a formal lab report that outlines all of the aspects of this lab exercise. Your formal lab report should include a title page, purpose, materials list, procedure, data gathered (tables and charts), and a conclusion.

Conclusions

1. What is the definition of an ecosystem?

2. Which of the two ecosystems you surveyed had the greatest amount of biological diversity (different number of species)?

3. Using the data on the biotic components of each ecosystem, which trophic level had the greatest abundance: the producers, consumers, or decomposers?

4. Describe the differences between the biotic or abiotic components you found in the disturbed and undisturbed ecosystems.

5. After performing this lab, what is the best way to identify whether an ecosystem has been disturbed by human activity?

LAB 51
Photosynthesis and Respiration

Purpose

In this lab, you will observe how plants undergo both photosynthesis and respiration, and will also identify the conditions in which these processes occur.

Materials

healthy Elodea plants
test tube racks
tape for labels
drinking straws
artificial light or window

clean test tubes with corks or caps (four for each group)
aluminum foil
glass jars or 500-ml beakers
pH indicator solution*

SAFETY CONCERN
WHEN PREPARING OR WORKING WITH THE pH INDICATOR SOLUTION, ALWAYS WEAR SAFETY GLASSES!

Procedure

Complete the following steps.

1. Each group should pour two jars or beakers one-half full with pH indicator solution. Charge one jar with carbon dioxide by blowing gently into the solution, using a straw as demonstrated by your instructor. When the solution is charged with enough carbon dioxide, it will turn to a yellow-green color. Label this jar with the tape "Carbon Dioxide Added (Yellow)." Label the other jar "No Carbon Dioxide (Blue)."

SAFETY CONCERN
BE CAREFUL WHEN BLOWING INTO THE STRAW. DO NOT SPRAY INDICATOR SOLUTION ALL OVER AS IT WILL STAIN YOUR CLOTHES AND SKIN.

---

*Before the laboratory is conducted, a stock solution of bromthymol blue must be made by mixing 0.5 grams of bromthymol blue powder with 500 ml of water. Once this is mixed, add 9 drops of sodium hydroxide (NaOH). Next, mix 20 ml of stock solution with 480 ml of water, and add 5 drops sodium hydroxide (NaOH). This is the solution that will be used to carry out the experiment.

2. Next, use the tape to label four test tubes A, B, C, and D, and place in a test tube rack.
3. Place a 4- or 5-inch section of healthy Elodea plants in test tubes A and C.
4. Carefully pour the Carbon Dioxide Added (Yellow) solution into test tubes A and B. Make sure to completely fill the test tubes and cap them.
5. Carefully pour the No Carbon Dioxide (Blue) solution into test tubes C and D. Make sure to completely fill the test tubes and then cap them.
6. Use aluminum foil to completely wrap test tubes C and D so as not to allow any light to enter.
7. Place all test tubes in a rack and put them in an area exposed to natural light.
8. Fill in Table 51–1 by placing a check mark for all of the correct parameters for each test tube, and let them sit undisturbed overnight.
9. After one day, observe the conditions in all four test tubes, and record the end color for each test tube on Table 51–1.
10. After you have recorded your results, clean up your experiment by following the specific clean-up procedures outlined by your instructor.

TABLE 51–1 End Color Chart								
	Light	No Light	CO_2	No CO_2	Plant	No Plant	Start Color	End Color
A								
B								
C								
D								

Conclusions

1. What occurred in the pH indicator solution to cause it to turn from blue to yellow when you blew into it through the straw?

2. Write the chemical reactions for both photosynthesis and respiration.

3. Which test tubes acted as controls in your experiment and why?

4. Which test tube underwent photosynthesis? Explain how your results proved this.

5. Which test tube underwent respiration? Explain how your results proved this.

6. Describe the specific conditions that cause plants to undergo photosynthesis and respiration.

LAB 52

Local Ecosystem Primary Production: How Much Does a Football Field Weigh?

Purpose

The purpose of this lab is to have you determine the productivity of a local ecosystem by calculating the amount of biomass it produces. You will then use your data to estimate the amount of biomass that a football field contains.

Materials

paper bags
balances or scales

string, 36 inches long

Procedure

Before you begin this activity, try and estimate how much all of the biomass in a football field weighs. Have your instructor take down each student group's guesses before beginning this lab exploration.

1. Go to the designated area outside of your school with your instructor to collect your biomass sample. You will be determining the productivity of a managed grassland ecosystem, so you need to collect a sample of grass. At the area designated by your instructor, select a section of healthy growing grass.
2. Surround the grass you have selected with your piece of string that has been shaped into a square roughly 9 inches long on each side.
3. Carefully remove all of the grass, including the root system from within your string square. Shake off as much soil from the roots as possible and place your biomass sample in a paper bag. Make sure to label your bag with your group's name.
4. You may now return to your classroom with your biomass sample. Because biomass is the total dry weight of plant material, you must let your sample dry out overnight. Place your bag, with the top open, in a warm sunny location in your classroom in order to dry it out completely.
5. After your biomass sample has been completely dried out, carefully determine its mass to the nearest tenth of a gram. Record your answer in Table 52–1. Get the data from two other groups in your class so you have the weight of at least three samples. Record these measurements in Table 52–1.
6. Convert all three masses of your samples into pounds by dividing your mass in grams by 453.6. Record your answers in Table 52–1.
7. Calculate the average for all three samples and record your answer in Table 52–1.

TABLE 52–1	Biomass Weights		
Sample #	Mass in Grams	Mass in Pounds	Calculations
Average			

Conclusions

1. Define the term *biomass*.

2. Using the data from Table 52–1, calculate how much biomass on average, one square yard of managed grassland weighs in pounds. Remember that your sample represents 81 square inches ($9'' \times 9''$), and one square yard equals 1,296 square inches ($36'' \times 36''$). Record your answer below. Show your work.

3. Next, you must calculate the area in square yards that a football field takes up. A football field is 110 yards from the back of one end zone to the back of the other end zone, and 53.3 yards wide. Record your answer below. Show your work.

4. Determine the weight of the football field by multiplying the weight of one square yard of your biomass by the the answer you calculated in Question 3. Record your answer below. Show your work.

5. To determine how close your "guess" was to the actual weight you calculated in Question 4, you must use the percent deviation calculation. The percent deviation of a measurement is determined by the difference between the actual value and your guess, divided by the actual value, then multiplied by 100. A percent deviation of 3 percent or less is considered accurate. Determine the percent deviation of your "guess" below. Show all of your work.

6. Briefly describe the process used to determine the productivity of an ecosystem.

LAB 53
The Effect of Nutrients on Plant Growth

Purpose

The purpose of this lab is to examine how a lack of specific mineral nutrients affects healthy plant growth. Because plants form the base of the food chain for all ecosystems, it is important to identify the specific role that certain abiotic minerals play in plant health. This lab also explores the concept of a limiting factor within an ecosystem.

Materials

 balance or scale
 2-liter soda bottles
 100-ml graduated cylinders
 stirring rod
 safety goggles
 plastic gloves
 plastic plant pots (three per student or group)
 perlite growing medium
 cucumber or bean seeds
 tape for labels
 digital camera (optional)
 plastic trays
 distilled water
 Macronutrient Stock Solutions:
　　1 M solution Ca $(NO_3)_2$
　　1 M solution KNO_3
　　1 M solution $MgSO_4$
　　0.5 M solution KH_2PO_4
　　0.5 M solution $CaCl_2$
　　0.5 M solution K_2SO_4
　　0.5 M solution KCl
　　Chelated iron solution (1 milliliter per liter)
 1 Liter of Micronutrient Stock Solution (add the following amounts to 1 liter of water):
　　boric acid (H_3BO_3) — 2.86 grams
　　manganous chloride $(MnCl_2)$ — 1.81 grams
　　zinc sulfate $(ZnSO_4)$ — 0.22 grams
　　copper sulfate $(CuSO_4)$ — 0.08 grams
　　molybdenum acid — 0.02 grams

Procedure A

1. Take three plastic pots and fill with perlite. Label one pot Nitrogen Deficiency, the second pot Phosphorus Deficiency, and the third pot No Deficiency.

2. Place your three pots on a plastic tray near a sunny window or under a light source. Add enough distilled water to each pot to moisten the perlite. Take three seeds of either beans or cucumbers and plant in the perlite about 1 inch deep. Keep the perlite moist, and wait three to five days for your seeds to germinate. After they have germinated, discard two of the seedlings from the pot. You only need to grow one plant per pot for the remainder of the experiment.

3. Next, make your three different nutrient solutions. Each group should share a 2-liter bottle of nutrient solution so that you do not make too much. Have your instructor help you decide which solutions your group should make. Use the information in Table 53–1 to mix a 2-liter batch of solution you need. Each group should have access to all three nutrient formulations to use to irrigate all three plants for the duration of the experiment.

4. After your seeds have germinated (approximately 1 week), begin to water them with the specific type of nutrient formula noted on their labels. Make sure that you do not overwater them! As the plants grow, observe and record their physical conditions in Table 53–2. If you have a digital camera available, take pictures of the different stages of growth to record the symptoms of nutrient deficiencies that appear during your experiment. Deficiencies should begin to appear within 2 to 3 weeks after germination.

5. Continue to water and observe your plants until they begin to flower and produce fruit.

Procedure B

1. After your experiment is complete, prepare a formal lab report presenting your investigation that includes the following: title page, purpose, procedure, data (tables and pictures), and conclusion.

TABLE 53–1

Nitrogen Deficiency Solution
(add to 2 liters of distilled H_2O)

$CaCl_2$	40 ml
K_2SO_4	40 ml
$MgSO_4$	8 ml
KH_2PO_4	8 ml
Iron chelate	8 ml
Micronutrient stock	4 ml

Phosphorus Deficiency Solution
(add to 2 liters of distilled H_2O)

$Ca(NO_3)_2$	20 ml
KNO_3	20 ml
$MgSO_4$	8 ml
KCl	4 ml
Iron chelate	8 ml
Micronutrient stock	4 ml

No Deficiency Solution
(add to 2 liters of H_2O)

$Ca(NO_3)_2$	20 ml
KNO_3	20 ml
$MgSO_4$	8 ml
KCl	8 ml
Iron chelate	8 ml
Micronutrient stock	4 ml

Conclusions

1. Explain why it was necessary to use distilled water when mixing your nutrient solutions.

2. Why could this experiment not have been performed by using soil instead of perlite?

3. What are the sources of nitrogen for plants in an ecosystem?

4. Describe the symptoms of nitrogen deficiency that you observed in your plants.

5. What are the sources of phosphorus for plants in an ecosystem?

TABLE 53–2					
Date	Plant Type	Height (cm)	Number of Leaves	Leaf Color	Other Characteristics

6. Describe the symptoms of phosphorus deficiency that you observed in your plants.

7. Explain how this experiment can be used to illustrate the concept of a limiting factor within an ecosystem.

LAB 54
Secondary Succession

Purpose

The purpose of this lab is to identify the changes that occur in the communities of an ecosystem during secondary succession. Biological succession is the slow, gradual change in the communities of a specific area, in a slow, predictable manner. During this lab your instructor will take you to a few different areas where different stages of secondary succession are occurring. The best location for this exercise is a wooded area near the edge of a farm field, although other successional areas may be used. The edge of the farm field represents a community in the early stages of succession, and the edge of the wooded area a middle stage of succession. The undisturbed interior of the woods represents a later stage of succession.

Materials

30-cm length of string
four 1-meter lengths of string
meter stick
thermometer
light meter

Procedure A

At each of the three locations, collect and record the following data:

1. Record the temperature of the soil using a thermometer that has its bulb inserted at least 2 inches into the ground.

2. Record the temperature of the air approximately 1 meter above the ground.

3. Record the light level directly at the ground surface.

4. Lay your 30-cm loop of string down on the ground, and record the number of the following type of green stem plants: narrow leafed, broad leafed over 50 cm, and broad leafed under 50 cm.

5. Next, lay out your four meter-length strings to form a square meter plot with your loop of string at its center. Remove the loop of string and record the number of the following type of woody stemmed plants within your square meter: shrubs (woody stemmed plants with two or more branches below 1 meter above the ground) and trees (woody stemmed plants with two or more branches above 1 meter above the ground).

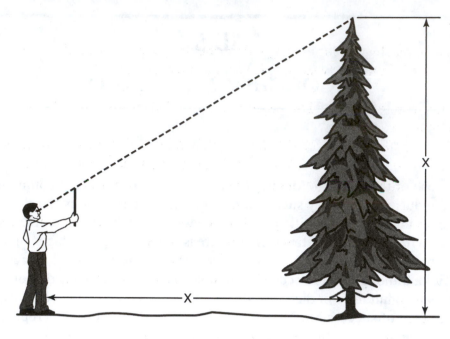

FIGURE 54–1

6. Record the height of the tallest plant within your square meter plot. If the tallest plant is too tall to measure with a meter stick, hold a meter stick at arm's length and close one eye. Carefully back away from the tree until the top of it lines up with the top of the meter stick. The distance you are from the base of the tree will be the approximate height of the tree (see Figure 54–1).

Procedure B

1. Use your data on the number of different plant types in each survey area to calculate the biotic abundance for each plant category using the following formula:

$$\text{biotic abundance (\%)} = \frac{\text{number of specific plant types in plot}}{\text{total number of all plants in plot}} \times 100$$

For example, if you counted 10 narrow-leaf plants in your plot, and the total number of plants within the plot is 100, then the biotic abundance of narrow-leaf plants is 10 percent.

2. Create a bar graph that compares the number of each plant category (narrow leaf, broad leaf below 50 cm, broad leaf above 50 cm, shrubs, and trees) for all three sites you surveyed. The x-axis should be labeled "Number of Plants," and the y-axis should be labeled "Plant Types." Make sure to make a key for your graph.

3. Prepare a formal lab report presenting your investigation that includes the following: title page, purpose, procedure, data (tables and charts), and conclusion.

Conclusions

1. Use the results of your survey to describe the unique abiotic characteristics of an ecosystem that is in the early stages of secondary succession.

2. Which type of plant had the highest biotic abundance in the early stage of succession for the plot you surveyed?

3. What are the unique biotic characteristics that you found for an ecosystem that is in the early stages of secondary succession?

4. Use the results of your survey to describe the unique abiotic characteristics of an ecosystem that is in the late stages of secondary succession.

5. Which type of plant had the highest biotic abundance in the latest stage of succession for the plot you surveyed?

6. What are the unique biotic characteristics you found for an ecosystem that is in the late stages of secondary succession?

LAB 55
Classification of the Living World
PowerPoint Presentation

Purpose

The purpose of this lab is to have you become familiar with the taxonomic classification of all living things on the Earth by creating your own PowerPoint presentation. You will also become familiar with the unique characteristics and examples of organisms that are classified within each of the five kingdoms used to classify organisms on Earth. If computers are not available to your class, you can create a poster or series of posters that present the characteristics and examples of organisms that make up the five taxonomic kingdoms. This can be done by using magazine clippings and original artwork, along with magic markers and colored pencils.

Materials

computer with Microsoft PowerPoint® application program
Internet access

Optional Materials

poster paper   magic markers
colored pencils   old magazines
glue sticks   scissors

Procedure

Complete the following steps.

1. Using your textbook, research the unique organisms that make up each of the five kingdoms. Make notes on the specific physical characteristics of the organisms that make up each specific kingdom. Also, identify examples of organisms that make up each specific kingdom. In some cases, you may wish to identify different phylums, classes, orders, or families of organisms within each kingdom. Record your notes in Table 55–1.

2. Next, you must collect images that will be used as visual examples for your presentation. You can search the Internet for the images of organisms you need for your presentation. Once you find a desirable image, right click on it with your mouse if you are using a PC, or click and hold on the image if you are using a Macintosh. You should then copy the image and paste it into a word processing file. Save all of your images in a word processing file for use later when you begin to create your presentation. Try to limit the amount of images to no more than twenty.

3. Once you have completed your research, you may begin to create a computer presentation using Microsoft PowerPoint. Have your instructor or a computer lab aide help you get started if you have never used PowerPoint before. There are many PowerPoint templates that can be used. Just make sure you pick a presentation type that uses both text and clip art. Try to limit the number of images you use for your presentation to twenty. This will ensure that your presentation will not use up a large amount of memory.

4. Finally, print out a hard copy of your presentation to hand in to your instructor.

TABLE 55–1 Organisms with the First Kingdoms		
Kingdom	Physical Characteristics	Example Phylum, Class, Order, Family, or Species

Conclusions

1. What are the five kingdoms used to classify organisms on Earth?

2. List the different physical characteristics that make up the organisms for each of the five kingdoms.

3. List two examples of organisms for each of the five kingdoms.